FIREF

GUIDE TO
FOSSILS

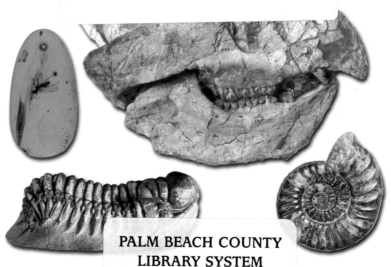

FIREFLY BOOKS

PICTURE CREDITS

Natural History Museum, London
5 Mawsonites
7 Oxynoticeras
10 Diatoms
11 Spider in Amber
23 Eryops
61 Raphidonema
63 Favosites - Frank Greenaway
65 Thamnopora
66 Lithostrotion
66 Lonsdaelia
69 Actinocyathus - Frank Greenaway
76 Fenestella
83 Silicified Brachiopod
84 Lingula
90 Productus - Frank Greenaway
93 Brasilia bradfordensis
102 Volutospina
106 Aturia - Frank Greenaway
106 Endoceras - Frank Greenaway
106 Eutrephoceras - Frank Greenaway
106 Orthoceras - Frank Greenaway
106 Nautilus
107 Model Ammonite
108 Goniatite
113 Parkinsonia
114 Hamites
117 Neohibolites - Frank Greenaway
117 Fissidentalium - Frank Greenaway
123 Modiolus - Frank Greenaway
125 Chlamys - Frank Greenaway
126 Cardiola
129 Dicranurus
129 Dalmanites
130 Phacops
130 Calymene
131 Calymene
132 Leonaspis
137 Hoploparis
137 Palaeocarpilius
138 Anthophorites
138 Acanthochirana
138 Snipe fly in amber
138 Libellulium - Frank Greenaway
145 Pentrimites
149 Micraster
151 Didymograptus
154 Coccosteus - Frank Greenaway
156 Cheiracanthus - Frank Greenaway
156 Osteolopis - Frank Greenaway
156 Lepidotus
158 Ichthyosaur
158 Turtle Eggs
158 Mesosaurus
159 Stenosaurus
159 Plesiosaurs
160 Tyrannosaurus
160 Protoceratops
161 Iguanodon
161 Pterosaur
165 Pliolophus
166 Zygolophodon
167 Homo Neanderthal
170 Annularia
171 Alethopteris
173 Araucaria
176 Acer Seeds
177 Anonaspermum - Frank Greenaway
177 Palaeowetherellia

Science Photo Library
4 Crinoid - Kaj R.Svensson
9 Tollund Man - Silkeborg Museum, Denmark/Munoz-Yague
12 Trilobite - Sinclair Stammers
14 Fossil Fish - Sinclair Stammers
18 Stromatolites - John Reader
43 Lucy - John Reader
51 Ammonites - Sinclair Stammers
58 Nummulites - Sinclair Stammers
116 Hoploscaphite - Peter Menzel
128 Trilobite - Sinclair Stammers
135 Trinucleus - Sinclair Stammers
137 Eurypterid - Kaj R. Svensson
145 Lapworthura - Martin Land
159 Dinosaur Eggs - Sinclair Stammers
162 Smilodon - Sinclair Stammers
171 Neuropteris - Sinclair Stammers
173 Gingko - Volker Steger
176 Platanus - Martin Land
182 Fossil forest - Sinclair Stammers
183 Trinucleus - Sinclair Stammers
184 Petrified Trees - Simon Fraser
187 Archaeopteryx - Jim Amos

A FIREFLY BOOK

Published by Firefly Books Ltd. 2003

Second printing 2004

National Library of Canada Cataloguing in Publication Data

Firefly guide to fossils. – 1st ed.

Includes index
ISBN 1-55297-812-5

1. Fossils – Identification.

QE714.F47 2003   560   C2003-900092–3

Publisher Cataloguing-in-Publication Data (U.S.)

Firefly guide to fossils. – 1st ed.
[192] p. ; col. ill. , maps : cm.
Includes index.
Summary: A guide to the identification, under-standing and hunting of fossils, which includes all major groups of fossils and an identification key.
ISBN: 1-55297-812-5
1. Fossils – Classification. 2. Fossils – Popular works. 3. Paleontology – Pictorial works. I. Title.
560/.22/2 21   QE714.F574 2003

Published in Canada in 2003 by
Firefly Books Ltd.
66 Leek Crescent
Richmond Hill, Ontario L4B 1H1

Published in the United States in 2003 by
Firefly Books (U.S.) Inc.
P.O. Box 1338, Ellicott Station
Buffalo, New York 14205

Published in Great Britain in 2003 by Philip's,
a division of Octopus Publishing Group Ltd,
2–4 Heron Quays, London E14 4JP

COMMISSIONING EDITOR Christian Humphries
EDITOR Joanna Potts
EXECUTIVE ART EDITOR Mike Brown
DESIGNER Alison Todd
PRODUCTION Sally Banner

Printed in China

COVER IMAGES
Front: Protoceratops skull and snipe fly in amber – Natural History Museum, London Trilobite – Science Photo Library/Sinclair Stammers
Back: Plesiosaurus – Natural History Museum, London

Some of the text originally appeared in Philip's Minerals, Rocks and Fossils, written by Dr W.R. Hamilton and updated by specialist contributors under the direction of Dr Brian Rosen.
Additional color photographs taken by The Photographic Unit of the Department of Exhibitions and Education, The Natural History Museum, London.

CONTENTS

INTRODUCTION

WHAT IS A FOSSIL?

Fossil crinoid or sea lily in a mudstone deposit. Thousands of extinct crinoid species have been found as fossils, but only a few hundred species of sea lily exist today. This crinoid is from the Silurian period.

Fossils are the remains of animals and plants more than 10,000 years old.

Fossils usually represent the **harder parts** of organisms, since these are the most resistant to decay and erosion. Most fossils therefore consist of the bones and shells of animals, or the leaves, seeds, and woody parts of plants.

Even the original hard parts may not be preserved intact, because biological processes like scavenging, and geological processes like waves and storms, tend to disarticulate, break and scatter all the various parts of the original organism. This is one reason why it is more common to find separate vertebrate bones than an entire skeleton, and individual crinoid plates are far more frequent than entire crinoids.

Another reason is that animals or plants may shed parts of themselves in the normal course of their life. For instance, leaves fall from trees, seeds float or are blown away from the parent plant, and many arthropods, like trilobites, molt their hard carapaces as they grow. **Trace fossils** are the footprints, burrows, impressions, and borings into rock left by organisms.

Fossils are found in the majority of **sedimentary rock** types. They are particularly common in limestones, marls, clays, siltstones, mudstones, and shales, and they are less common in sandstones, conglomerates, and graywackes. Fossils can also be preserved in sedimentary ironstones and in volcanic ash.

The majority of fossils are **aquatic** animals or plants (i.e. they lived in seas, rivers, lakes, or estuaries), because conditions for preservation are usually better in aquatic environments than on land. Even **terrestrial** animals and plants are more likely to be preserved in aquatic sediments, either through drowning (in a flood, for example), or because they fell into water, or were swept into it by floods and other sudden events. In this way, fossil land mammals are often found in the same deposits as the remains of fish, crocodiles, or turtles.

Occasionally, entire organisms are preserved in frozen soil (such as mammoths), peat bogs, and asphalt lakes, or trapped in hardened resin (such as insects in amber).

FOSSIL FORMATION

Fossils represent only a tiny fraction of the total number of animals and plants that have existed, because there are several conditions that must be fulfilled before fossilization is possible. The chances that an organism will be fossilized are greatly increased if it has hard parts, such as a shell, bones, or a test.

Cast of the fossil jellyfish Mawsonites spriggi from the Precambrian Ediacara fauna of South Australia. The Ediacara fauna is a unique fossil deposit, famous for its preservation of rare, soft-bodied organisms.

The ability of an organism to survive destructive surroundings often depends on the size of the animal or plant, and the strength or percentage of hard parts. For this reason, a large **mollusk** is more likely to survive than a tiny, more fragile shell. Even so the future of an organism can depend on the speed with which it becomes buried in the sedimentary soil. Given the right environmental conditions, a soft-bodied **jellyfish** may be preserved quickly enough to leave an imprint in the soil, despite the fleshy nature of its body. These types of soft body fossils are extremely rare.

Once a fossil has been formed, conditions within the rock itself can continue to threaten the fossil's preservation. In the course of the millions of years the fossil may have been encased in rock, gigantic changes will have taken place in the Earth's crust. Changes in the form and structure of the rock itself may have forced the shape of the fossil to change, or may have crushed it beyond recognition.

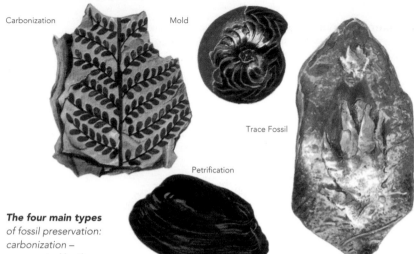

Carbonization

Mold

Trace Fossil

Petrification

The four main types of fossil preservation: carbonization – represented by the coal measure plant Odontopteris; *internal mold* – the ammonite Goniatite; *trace fossil* – a dinosaur footprint; and petrification – depicted by the bivalve Carbonicola.

Unaltered Hard Parts

The harder parts of animals and plants contain a number of materials that do not decay, such as **phosphates** in bone and **calcium carbonate** in shells. Among **invertebrate** animals, the shells of clams and the hard, segmented bodies of insects are frequently found unchanged in rocks less than 60 million years old. In these instances, the original structure of the hard parts and their chemical composition bear comparison with living relatives. Hard parts, composed of **calcium phosphate**, are resistant to chemical change, and the remains of various animals may be found unaltered in rocks more than 60 million years old.

Altered Hard Parts

The shells of various **invertebrate** animals are made up of minerals possessing a distinctive fibrous or layered structure. The retention of this structure indicates that the hard parts are probably unaltered. A mosaic or granular interlocking texture will suggest **recrystallization**. This process is the result of the original fibrous material going into solution and reforming with a coarser growth structure.

Replacement, on the other hand, infers that the original material of the fossil is replaced with another. During fossil formation, minerals are often replaced by others that are better able to survive the rigors of the subterranean environment. For example, **pyrite**, **hematite**, or **quartz** commonly replaces minerals in a shell or a bone, molecule by molecule, so a **calcium-based** coral may be preserved as a hematite fossil. Pyritized fossils are often found in **black mudstones** (fine-grained rocks deposited in environments where oxygen is lacking).

Silica is another common replacement mineral. **Quartz** or **chalcedony** will often replace **calcite** in shells or in the skeleton of corals. This can be observed when such fossils protude from a limestone surface. Silica is also known to impregnate **fossil wood**. Mineral-rich water seeps into the empty spaces of dead, buried trees or animals and here the delicate tissues are petrified or literally turned to stone. This process is called **petrification**. Petrification makes the fossil denser and heavier, but it retains its shape, and in the case of trees can still show the growth rings. The **Petrified Forest National Park** in Arizona, United States, has many examples of petrified wood.

Generally wood and plant materials are rich in volatile substances. The same is true of some invertebrate skeletons composed of organic materials, and the process of **carbonization** can affect both. Carbonization produces a decrease in the original volatile substances such as oxygen, hydrogen, and nitrogen, and preserves the leaf or skeleton as a thin film of **carbon** lying along the bedding plane of a sedimentary rock. The layer of carbon demonstrates the appearance of the original organism.

Internal and External Molds

The organic part of a creature is always the first to decay. This leaves a gap in the shell that is often filled with soft sediment. The shell is more soluble than the rock, and slowly dissolves leaving a sedimentary-rock cast of the inside of the shell. Such cores are called **internal molds** or **steinkerns**. In the case of clams, the steinkern will not take the internal form of the animal, and you may be able to recognize the size and position of the muscle scars and perhaps the area occupied by the internal organs.

Under the right circumstances, the shell material is replaced by some other substance – often a type of silica – to give a cast of the shell's exterior. The 'replacement' may even be the original shell material, recrystallized.

Fossil ammonite
Oxynoticeras oxynotum *with distinctive green calcite chambers and yellow pyritized septa. Pyrite and calcite are common replacement minerals in fossils.*

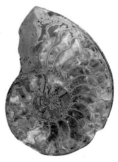

Sometimes no sediment fills the empty shell. The solutions that permeate the strata may totally dissolve a buried shell, leaving a mold behind in the rock. This is a common phenomenon in sandstones, sandy limestones, or ironstones, the cavities being called **external molds**. The tracks of living creatures can also be fossilized, and these are known as **trace fossils**.

Fossils A *The organic part decays first, leaving a gap which is filled with sediment [1]. The shell slowly dissolves [2], leaving a rock cast of the inside [3]. The shell may be replaced by another material, giving a cast of the exterior [4]. Sometimes no sediment fills the shell [5]. Solutions may permeate [6], causing the shell to dissolve leaving a mold in the rock [7]. A replica may later be cast [8].*

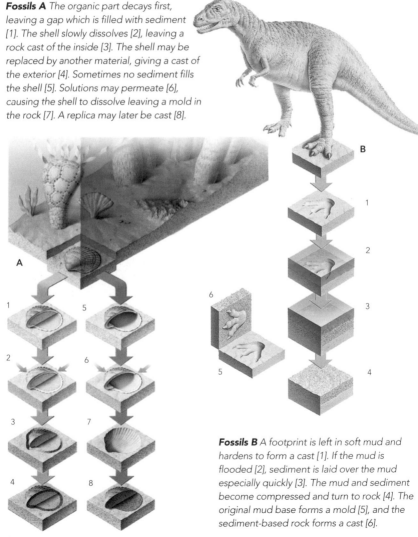

Fossils B *A footprint is left in soft mud and hardens to form a cast [1]. If the mud is flooded [2], sediment is laid over the mud especially quickly [3]. The mud and sediment become compressed and turn to rock [4]. The original mud base forms a mold [5], and the sediment-based rock forms a cast [6].*

FOSSIL DEFINITIONS

Most of the fossils you will find represent the hard parts of an organism, such as a shell, tooth, or bone. Other fossil types, including casts, molds and impressions, are of equal importance in providing information on individual organisms or even **fossil communities**. Shells and skeletons are frequently termed **body fossils**, because they provide details on the shape and functions of the actual organism. The same may be said for an impression – particularly when it retains details of the soft parts. Body fossils occur in all shapes and sizes, ranging from microscopic sea-dwellers to huge terrestrial **dinosaurs**. Their preservation will vary according to the conditions that prevailed at the time of death and burial.

Mummified body of Tollund Man, a 'bog body' (220–240 BC) found in Jutland, Denmark. The well-preserved body cannot be regarded as a true fossil since it is less than 10,000 years old. Rather it is an excellent example of a subfossil. Low temperature and oxygen, and high tannins, led to his mummification. The man was hanged by a leather noose.

Subfossil
This is a term sometimes applied to the remains of animals and plants preserved in rocks less than 10,000 years old. These include the remains of bison trapped in peat bogs, or of ancient man mummified in caves. Subfossils were formed after the last Ice Age, during the **Holocene** epoch.

Microfossils
These are usually less than 1/50 inch (0.5 millimeters) in size, but organisms regarded as microscopic can deposit skeletons up to four inches (10 centimeters) in diameter. Both single-celled plants and animals can form mineralized skeletons, and some make a major contribution to the formation of sedimentary rocks.

Macrofossils
These are greater than 2/5 inch (one centimeter) in size. The term is usually applied to the more advanced plants and animals, such as clams, corals, or the skeletons of vertebrates.

Unusual fossils
These are, by definition, extremely rare. They include mammoths dug from the Siberian wastes and the remains of the first bird Archaeopteryx. The term 'unusual' refers to the mode of preservation, in which a combination of events and conditions results in all or most of the organism being preserved in the rock.

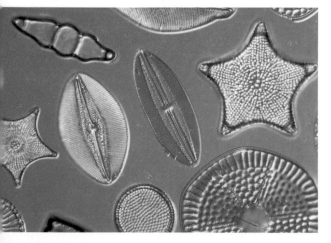

Fossil diatoms on a slide, viewed under a light microscope using differential interference contrast. Fossil diatoms are an example of microfossils – fossils less than 1/50 inch (0.5 millimeters) in size. Diatoms are unicellular algae and have an extensive fossil record going back to the Cretaceous (142–65 million years ago). Some rock types consist almost entirely of fossil diatoms.

Famous deposits include the **Solnhofen Limestone** of southern Germany and the **Burgess Shale** of Canada.

Some of the most spectacular fossils result from preservation in special conditions. For example, **amber** is fossil resin and occurs in deposits around the Baltic Sea. When the deposits of amber were accumulating, insects and spiders occasionally became trapped in the sticky resin and their complete bodies and even colors have been preserved for millions of years.

In the northern parts of Asia and North America, the soil has been permanently frozen for several thousands of years (permafrost) and complete mammoths, rhinoceroses, and other mammals have been discovered preserved in these frozen deposits. Often these remains are so well preserved that even the stomach contents can be studied, providing vital clues to the lifestyle of many extinct animals.

A fossilized spider, Abliguritor niger, *captured in amber from the Baltic during the* Oligocene epoch (34–44 million years ago). *Amber is an excellent preserver, and trapped insects are often fossilized complete. Animals preserved in this way are 'unusual fossils.'*

Trace fossils

These are formed by organisms performing the functions of everyday life, such as walking, crawling, burrowing, boring, or simply feeding. Dinosaur footprints, worm trails, and clam burrows are all trace fossils. These characteristic traces sometimes reveal the presence of animals that have not been preserved in any other form.

Coprolites

These are also **trace fossils.** They are the preserved droppings of animals. They can vary in size from the tiny fecal pellets of a sea-snail to the large coprolites of crocodiles, dinosaurs, or mammals. Coprolites can be useful in determining the eating habits of past animals and often contain the fossilized remains of organisms that have not previously been found.

Bioclasts

These are fossils or fragments of fossils enclosed in sediments. Bioclasts usually are hand specimens or thin sections viewed under the microscope.

Non-fossils

Despite their age, the ancient remains of man-made tools and other inorganic structures are not defined as fossils.

GEOLOGY

THE DISTRIBUTION OF FOSSILS IN TIME AND SPACE

Geology is the study of the materials of the Earth, their origin, arrangement, classification, change, and history. **Paleontology**, the study of fossilized plant and animal remains, is elemental to the determination of the Earth's geological history.

In much geological work, fossils are used simply as **markers**, which indicate the age of the rocks in which they occur. When **sedimentary rocks** are deposited, the oldest are at the bottom, and the layers of rock become progressively younger as we move up the sequence. Unfortunately this initially simple situation may become very complicated as a result of different rates of deposition or total breaks in deposition, erosion of the surface, or folding and faulting of the rocks. The presence of specific fossils can help to establish the correct sequence of rock layers, or strata. Sequences of rock that share specific characteristics such as grain size and/or sedimentary structures are known as **facies**. Some fossils may be restricted to given facies, others to a specific horizon or time zone. They may be defined as follows.

FOSSIL DISTRIBUTION – DEFINITIONS

Facies fossils

These are restricted to sediments that have given

Trilobites make good zone fossils. Shown are the eyes of Trevoropyge prorotundifrons, which lived from the Lower Ordovician to the Upper Devonian (495–350 million years ago). It had one of the first advanced visual systems in the animal kingdom. Its large eyes consisted of about 700 separate lenses, which gave it an excellent field of vision. It was found in Morocco.

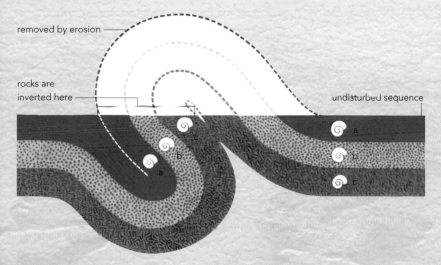

removed by erosion

rocks are
inverted here

undisturbed sequence

a
b
c

a
b
c

characteristics. Facies fossils are of little or no use to the stratigrapher because they may be limited in terms of geographical distribution. However, they may be useful in the interpretation of environments.

Zone fossils

These are restricted to specific levels within the **stratigraphic column**. These levels are called **biostratigraphic units**, and may be defined by the range of a given fossil, or in the case of an **assemblage zone** by total fossil content. The term is synonymous with **index fossil**. The most useful zone fossils are short-lived and widespread in their distribution. **Graptolites**, **ammonites**, **foraminiferids**, and **trilobites** change quickly with time and are good zone fossils.

Range

This is used to describe the time interval during which a fossil existed on Earth. This time interval is measured in terms of **rock units**, which may be part of a period, or may possibly span several periods of geological time.

Derived fossils

These are remains that have been eroded from one bed, and then were transported and deposited in a younger (more recently formed) bed. They are therefore older than the sediment in which they are enclosed when discovered.

Folds in the rock strata *can be detected by the presence of certain zone fossils. A geologist first needs to study an undisturbed section of rock in order to assess the correct sequence of rocks and their associated zone fossils. This will enable a geologist to discover whether the rocks in adjacent areas have been folded or not.*

13

Faunal provinces

These define the distribution of associated groups of organisms. The boundary of the province may be marked by climatic or geographic criteria. In the **stratigraphic record**, it is almost impossible to recognize these criteria with any degree of accuracy. A province is therefore identified by the association of fossils in one area at a particular interval in the geological timescale. The association may be referred to as a **faunal assemblage**. Provinces may vary in terms of their spatial distribution. An accurate reconstruction of a faunal province will probably provide information on the Earth's plate movements and the **paleogeography** of the period.

Life assemblage

This is used to describe a group of organisms that have remained **in situ** after death. Reefs are good examples of life assemblages.

Derived assemblages

These are accumulations of organisms that have been transported and subsequently buried outside their normal habitat or living area.

Shoal of fossilized fish (Knightia alta) *from the Eocene period (55–34 million years ago). This well-preserved fossil is called a life assemblage because it shows a group of animals that were killed suddenly and then preserved where they lay. The fish are teleosts (bony fish) and were found in the Green River Formation in Kemmerer, Wyoming, United States.*

FOSSIL COMMUNITIES

For the biologist, a community study will involve an analysis of the interrelationships that exist between the various organisms that comprise the community.

For the paleontologist, this type of study is restricted by the disappearance of most of the evidence. For the paleontologist, a community is best defined as a group of organisms that lived in the same habitat. The paleontologist can analyze the structure and function of a shell or skeleton, and suggest modes of life for individual species. It is also possible to recreate the general structure of a community. An obvious example is a fossil reef where the paleontologist could rely on a modern-day equivalent for essential information. The study of an individual fossil in relation to its habitat is termed **paleoautecology**, while the study of groups or assemblages of fossils is termed **paleosynecology**.

A truly meaningful analysis of a community is restricted by the fact that the **fossil record** is incomplete, and that 70 percent or more of an original marine community may have been soft-bodied. Nevertheless, they represent a true, albeit incomplete, record of a fossil community.

The discovery of a highly fossiliferous bed (such as the sea-floor fossils of the **Wenlock series**) may lead to the description of a fossil community. It is essential, therefore, to have some knowledge of the fossils that may occur in the same place at the same time. **Precambrian** communities are very rare, but from the **Cambrian** onward the record of recurring assemblages greatly improves.

GEOLOGICAL TIMESCALE

Planet Earth is approximately 4.6 billion years old. This age is established by using dating techniques – called **radiometric dating** – that measure the rate of radioactive decay in given rocks. Geological time is split into several major divisions or **eras** of which the last three – **Paleozoic**, **Mesozoic**, and **Cenozoic** – contain the vast majority of known fossils. The eras are divided into eleven major **periods**, which are themselves subdivided into epochs.

Of the complete span of geological time, the **Precambrian** lasted slightly more than 4 billion years. Precambrian rocks are commonplace in continental shield areas, such as the Canadian Shield. They are mostly altered or metamorphosed by heat or pressure. Fossils are rare in these rocks – the first communities appeared about 670 million years ago. The 600 million years from the Precambrian to the present day are marked by the

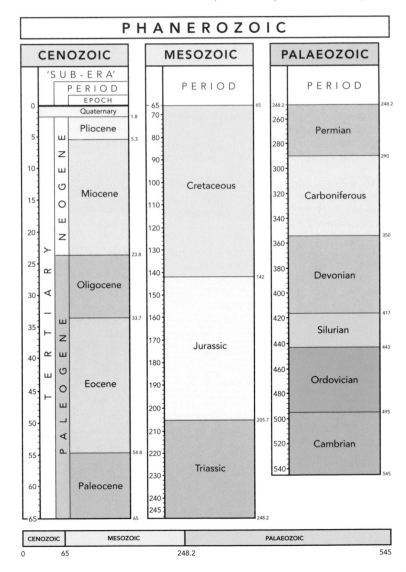

presence of various forms of life. The term **Phanerozoic** meaning 'obvious life' applies to this span of geological time. The superimposition of sedimentary layers is another feature of this time span, and the evolution and extinction of many organisms provide a key to the subdivision of the Phanerozoic. Subdivision by the grouping of layered rock sequences is called **lithostratigraphy**, whereas grouping by the use of fossils is known as **biostratigraphy**.

The terms **Lower, Middle,** and **Upper** are used to refer to rocks within a period or era; however, the terms are not interchangeable. For example, it is correct to say that the earliest mammals occur in the **Upper Triassic** – i.e., in Upper Triassic rocks – or that the mammals originated in Late Triassic times, but it is not correct to say that the mammals originated in Upper Triassic times.

THE TWELVE MAJOR PERIODS

Precambrian

The discovery of fossils in sediments of Precambrian age is a rare occurrence. Algal **stromatolites** are the most common fossils found in Precambrian deposits. The oldest appeared 3.35 billion years ago. Fossil animals appeared much later, and the record of the **first true fauna** occurs in rocks between 680 and 600 million years old. This fauna was first discovered in

Charnia

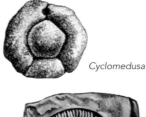

Cyclomedusa

worms

Precambrian fossils finds are very rare. The first worms appeared about 650 million years ago. They were the ancestors of many advanced groups of mammals. Planktonic jellyfish Cyclomedusa is probably the most common and widespread Precambrian fossil. The colonial organism Charnia reaches up to 3.3 feet (one meter) in length. It was first found in Charnwood Forest, central England.

17

the Pound Quartzite of **Ediacara**, South Australia. It contains a number of soft-bodied jellyfish, hydrozoans, and worms. Initially the fauna was thought to be confined in space and time, but further occurrences have now been recorded in Siberia, Europe, Canada, and south west Africa.

It is possible that the Ediacara fauna or 'Vendian biota' is also our first true fossil community.

Cambrian

Earliest period of the **Paleozoic** era, lasting from c.545 million to 495 million years ago. The rocks of this period are the earliest to preserve the hard parts

Precambrian stromatolites at Shark Bay, Western Australia. Stromatolites consist of many cyanobacteria. These masses are among the oldest organic remains to have been found, ranging from 2,000 to 3,000 million years old.

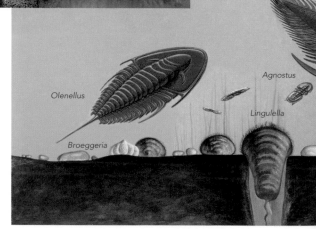

The Cambrian period was dominated by trilobites. In the Lower Cambrian, Olenellus was common in north-west Scotland, Greenland, and parts of western Canada and the United States. Paradoxides is a Middle Cambrian trilobite. It is often found with the brachiopod Lingulella and the gastropod-like Hyolithes.

of animals as fossils. Cambrian rocks contain a large variety of fossils, including all the animal phyla with the exception of the vertebrates. During the Cambrian period, the animals lived in the **seas** and the land was barren. The most common animal forms were **trilobites, brachiopods, sponges,** and **snails**. The diversity of trilobite species during this period suggests a long period of evolution beforehand. Cambrian trilobites are good **zone fossils**, and vary significantly depending on the faunal province. Plant life in the Cambrian period consisted mainly of seaweed.

Ordovician
Second-oldest period of the **Paleozoic** era, from 495 to 443 million years ago. All animal life was still restricted to the sea. Numerous invertebrates flourished, such as **trilobites, brachiopods, corals, graptolites, mollusks,** and **echinoderms**. The only

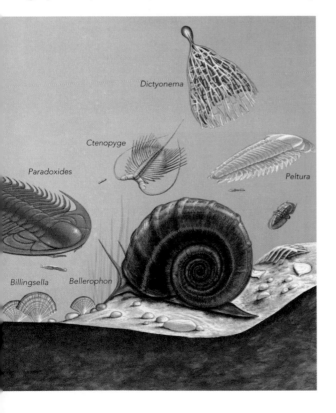

Orthoceras

Dalmanella

Orthambonites

Lophospira

Nautiloids, *a distant relative of the squid, were the major scavengers and carnivores of the Ordovician period. Trilobites and brachiopods dominated the shallow water communities. The rapid increase in the number and variety of graptolites make them an ideal zone fossil.*

major Cambrian group not present in the Ordovician period was the reef-building **archaeocyathids**, suggesting a minor ecological crisis at the end of the Cambrian. Trilobites and brachiopods continued to dominate the shallow-water communities. Some Ordovician trilobites developed defensive tail-tucking mechanisms. The size and thickness of brachiopod shells throughout the Ordovician period can help to determine the environment and lifestyle of an individual. The **nautiloids** (a kind of pre-squid carnivore) were successful during this period and make ideal **zone fossils**. Remains of **jawless fish** in coastal deposits mark the first record of the vertebrates.

Silurian

Period of the **Paleozoic** era, lasting from 443 to 417 million years ago. **Brachiopods, corals, bryozoans,** and **crinoids,** often termed the 'shelly fauna' flourished in shallow waters. The richly fossiliferous bedding planes of the **Wenlock limestone** show the ancient sea floor, with organisms often preserved in situ. **Graptolites** existed in the deeper waters. Marine invertebrates resembled those of **Ordovician** times, and fragmentary remains show that **jawless fishes** (agnathans) began to evolve. Their main

enemies were the **eurypterids,** a form of giant arthropod. The earliest **land plants** (psilopsids) and first **land animals** (archaic mites, millipedes and some scorpions) developed. Mountains formed in north west Europe and Greenland.

Wenlock Limestone *has many shallow water fossils from the Silurian. Leptaena and* Atrypa *brachiopods were common.*

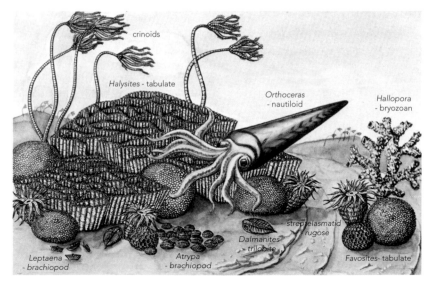

21

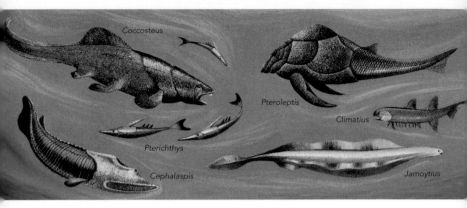

Coccosteus
Pteroleptis
Climatius
Pterichthys
Cephalaspis
Jamoytius

The Devonian is also known as the 'Age of Fishes.' The heavily armored jawless Cephalaspis *is perhaps the smallest advance from the crustacean. Next in the evolutionary chain came* Coccosteus, *whose mouth opened vertically.* Pterichthys *had long and narrow wing-like appendages, which presumably it used in self-defense.* Pterolepis *was an active swimmer whose tail tipped downward, the reverse of today's shark.* Climatius *was an acanthodian or 'spiny shark.'* Jamoytius *was an anaspid with elongated paired fins.*

Devonian

Period of the **Paleozoic** era, lasting from 408 to 360 million years ago. It is sometimes called the '**Age of Fishes.**' Numerous marine and freshwater remains include jawless fishes and forerunners of today's fish. At first, **heavily armored** fish such as *Cephalaspis* and *Coccosteus* dominated, but by the end of the period these had been replaced by sharks, lungfish, and ray-finned bony fish. **Brachiopods, corals, bryozoans,** and **crinoids** were also common. The **first** known **land vertebrate**, the amphibian **Ichthyostega**, appeared at this time. Land animals included scorpions, mites, spiders, and the first insects. Land plants consisted of tall club mosses, horsetails, and ferns. In Devonian times, much of the British Isles was desert mountain environment or semi-desert coastal plains, giving rise to the red rock known as **Old Red Sandstone**.

Carboniferous

Fifth period of the **Paleozoic** era, lasting from 350 to 290 million years ago. The Carboniferous divides into two series. In North America, the Lower Carboniferous is known as the **Mississippian period** and the Upper Carboniferous as the **Pennsylvanian period**. The **Lower Carboniferous** had shallow, warm seas and marine limestone with a coal-rich fauna. **Corals** and **brachiopods** were numerous. The size and number of fossils suggest that the warm waters offered ideal conditions. The **Upper Carboniferous** is dominated by river and deltaic sediments containing coal seams formed from swampy forests of conifers and tree ferns. Land was often flooded and plants alternate with non-marine **bivalves**. Amphibians adapted and prospered.

Permian

Last period of the **Paleozoic** era, lasting from 290 to 248 million years ago. There was widespread geologic uplift, resulting in the formation of Pangaea – a single supercontinent. The major climatic characteristics of the Permian were aridity and glaciation. These changes caused the **extinction** of numerous marine invertebrate animals including the trilobites, rugose and tabulate corals, and many brachiopod families. Climatic change meant that **reptiles**, such as *Eryops*, thrived.

Eryops

Triassic

First period of the **Mesozoic** era, lasting from 248 to 206 million years ago. Following a wave of extinction at the close of the **Permian** period, many new kinds of animals developed. New groups began to replace

The Carboniferous period saw the emergence of swamps and vast forests with gigantic 'scale trees,' such as Lepidodendron and horsetails, such as Sphenophyllum. Amphibians, such as Eogyrinus, thrived. Cordaites are the ancestors of the true conifer. This period saw the accumulation of large coal reserves.

Cordaites

Meganeura

Sphenophyllum

Eogyrinus

Hylonomus

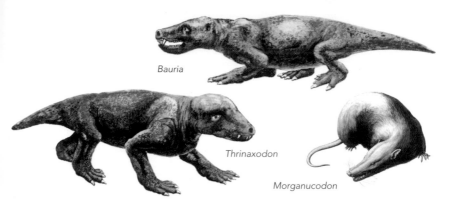

Bauria

Thrinaxodon

Morganucodon

Reptiles such as Bauria ruled the Early Triassic. Bauria was a specialized herbivore. The Late Triassic saw the transition from reptiles to the first mammals. Thrinaxodon was a small cynodont, or mammal-like reptile about the size of a cat. Morganucodon was a small mammal.

previously dominant ones; for example **ammonites** developed and replaced the **goniatites**, while **bivalves** began to rival the dominant **brachiopods**. On land lived the first ancestors of the **dinosaurs** – the **thecodontians**. Mammal-like reptiles were common, and by the end of the Triassic period the first true **mammals** existed. In the seas lived **placodonts**, **nothosaurs** and the **Ichthyosaurs**, fast-swimming, predatory marine reptiles with a dolphin-shaped body. The first frogs, turtles, crocodilians, and lizards also appeared. Plant life consisted mainly of primitive non-flowering plants, with ferns and conifers predominating.

Jurassic

Central period of the **Mesozoic** era; it lasted from 206 to 142 million years ago. The **Tethys Sea** divided the supercontinent **Pangaea**. Shallow seas covered much of Europe and **bivalves** flourished. Non-flowering flora was dominant. **Dinosaurs** were now the ruling reptiles. In this period there were large **saurischian** dinosaurs, such as **Atlantosaurus** and **Stegosaurus** and the **Megalosaurus–Allosaurus** group produced the major predators. **Archaeopteryx**, **plesiosaurs**, and **pterosaurs** also lived in the Jurassic.

Diplodocus is the largest animal that has ever walked the Earth, 90 feet (27 meters) long. It lived during the Jurassic period, in what is now the northern United States. It was a swamp-dwelling herbivore.

Styracosaurus

Protoceratops

Cretaceous

Last period of the **Mesozoic** era, lasting from 142 to 65 million years ago. The shelf seas continued to withdraw, a process that started at the end of the **Jurassic** period. The Upper Cretaceous marked a rapid diversification of the **ornithischians** with numerous species of duck-billed or crested, horned, and heavily armored families living in herds in North America and Asia. **Dinosaurs** flourished until the end of the period, when they died out in a **mass extinction** that killed off a vast range of vertebrates and invertebrates. Controversy surrounds the reasons for this mass extinction, which, in terms of the number of species affected, was actually smaller than the event that brought the **trilobites** to extinction. The Cretaceous period also saw the first true placental and marsupial **mammals** and flowering plants appear. The chalk rocks of north west Europe were deposited during the Upper Cretaceous.

Ceratopsians were rhinoceros-like, plant-eating dinosaurs of the Late Cretaceous period. They were one of the last major groups of dinosaurs to evolve. Styracosaurus was named in 1913 from a fossil found in Alberta, Canada. Protoceratops was first discovered in the 1920s in the Gobi Desert, central Asia.

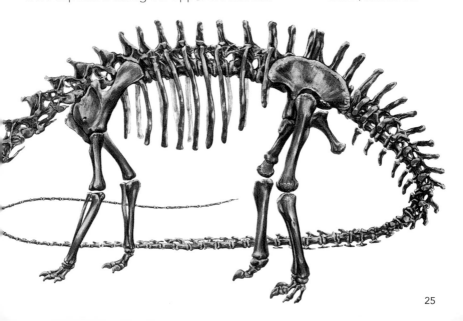

The Cenozoic is the 'Age of Mammals.' The opposum-like Plesiadapis *are early primates.* Pliolophus *is the earliest ancestor of the modern horse. Traces of Paleocene oak, elm, poplar, redwood, cypress, and yew trees have been found at Menat, central France.*

Paleogene and Neogene

The **Tertiary** 'sub-era' of the **Cenozoic** era lasted from 65 million to 2 million years ago. It divides into two periods: Paleogene and Neogene. The Paleogene includes the **Paleocene**, **Eocene**, and **Oligocene** epochs. The Neogene period includes the **Miocene** and **Pliocene**. Early Tertiary times were marked by great **mountain-building** activity (the Rockies, Andes, Alps, and Himalayas) and continents began to resemble our own. Seasons began to be more distinguishable and animals began to practice migration. The plant world, although dominated by **angiosperms**, began to develop species similar to those of the present day. Both marsupial and placental **mammals** diversified greatly. Archaic forms of carnivores and herbivores flourished, along with early primates, bats, rodents, and whales.

Quaternary

Most recent period of the **Cenozoic** era, beginning about 2 million years ago and extending to the present day. It divides into the **Pleistocene** epoch, characterized by a periodic succession of great **Ice Ages**, and the **Holocene** epoch, which started some 10,000 years ago. Climate change led to variations in plant and animal communities. The preservation of glacial communities is generally poorer than that of warmer ones, so detailed evidence regarding their lives and environment is rare. Genetic analysis has shown that all people living on Earth today belong to the human species **Homo sapiens**, which first evolved in Africa c.150,000 years ago. There are also what are known as archaic *Homo sapiens* fossils, which date back to 250,000 years ago in Africa. No continuous record exists between the two.

EARTH PROCESSES

If you look at a map of the world today, it is interesting to note the corresponding shapes of various continents. If the shape of the continental shelves – the gradual sloping of a continent under the water – was included in the map, then the fit between the continents would appear even closer. This suggests that the continents were at some previous time, joined together.

In 1912 German scientist **Alfred Wegener** first proposed the theory of **continental drift**, but the mechanism for moving the continents remained a mystery to him. **Plate tectonics**, the idea that the continents ride on a series of rigid but highly mobile plates, has since provided an answer.

Below the Earth's plates, powerful temperature differences create massive, slow movements of molten and plastic rock. These streams, known as **convection currents**, circulate towards the surface of the mantle. There are weak points at the plate margins, usually at the mid-ocean ridges where the crust is much thinner. Here, two currents collide and separate and the plates bulge and move apart. Magma – molten rock from the mantle – flows up to the surface and solidifies. It pushes apart the relatively mobile **ocean crust** and, if there are no intervening subduction zones, pushes the continents with them. Rocks on the sea bed grow progressively younger the nearer they are to the ridge. **Subduction** is the plunging of heavy oceanic crust beneath the lighter **continental crust**. It results in chains of magmatic volcanoes. Subduction zones exist under such diverse landscapes as the Aleutian Islands and the Andes mountains.

Continental drift
About 180 million years ago, the original Pangaea land mass began to split into two continental groups, which further separated over time to produce the present-day configuration.

180 million years ago

135 million years ago

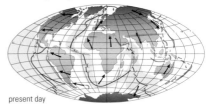
present day

27

The presence of identical fossil species on the matching coasts of South America and Africa was the most compelling evidence supporting Wegener's theory of continental drift. It was physically impossible for fossils to have traveled or been transported across the vast Atlantic Ocean, and thus the two continents were once joined. Also, the discovery of fossils of tropical plants (in the form of coal deposits) in Antarctica led to the conclusion that this frozen continent must once have been situated closer to the equator, in a more temperate climate where lush, swampy vegetation grew.

Ordovician The most significant movement of the plates during the Ordovician period was the movement of North America towards northern Europe, thus compressing the sea area between the two land masses.

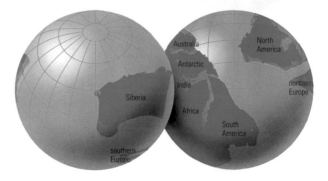

Silurian During the Silurian period the distance between the North American and northern European land masses greatly reduced. By this time all the southern continents had fused together, forming the land mass known as Gondwanaland.

Devonian By the Devonian period, North America had collided with northern Europe, and the sediments between them had been thrust up to form the Caledonian Mountains.

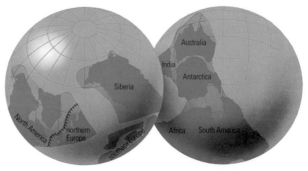

28

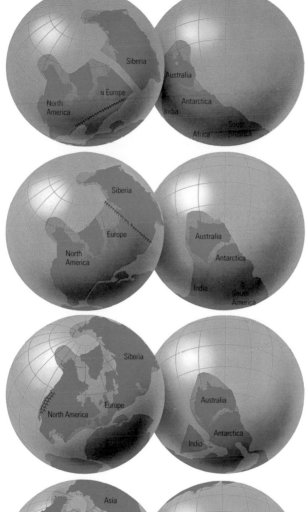

Permian During the Permian period northern Europe collided with southern Europe, pushing up the Variscan-type fold mountains. This combined block began to move toward the Siberian plate.

Triassic During the Triassic period, the collision between the North America-Europe landmass and the Siberian plate pushed up the Ural Mountains.

Jurassic By the Jurassic period, the Gondwanaland landmass had begun to break up into separate continents.

Paleocene During the Paleocene epoch, the movement between Africa and Europe raised the Alps. A great area of volcanic activity reached from the British Isles toward the position of Iceland.

Pangaea

Scientists now believe that a single supercontinent – **Pangaea** – was created in the **Permian** period and remained whole until the middle of the **Mesozoic** era. In the **Jurassic** period, this landmass split into two separate continents, now known as **Laurasia** (northern Pangaea) and **Gondwanaland** (southern Pangaea), and separated by the **Tethys Sea**. During the **Cretaceous** period, the present-day continents of North America and Europe began to break away from Laurasia, thereby forming a small predecessor of the Atlantic Ocean. The land masses further separated over time to produce the present-day configuration.

The discovery that the continents shift along on the top of slowly moving crustal plates provided support for the theories of continental drift. The plates converge and diverge along margins marked by seismic and volcanic activity. Plates diverge from mid-ocean ridges where molten lava pushes up and forces the plates apart at a rate of up to 1.5 inches (3.75 centimeters) a year. Converging plates form either a trench (where the oceanic plates sink below the lighter continental rock) or mountain ranges (where two continents collide).

Fossils – evidence of continental drift

Biological clues support the geological evidence for continental drift. *Glossopteris*, a deciduous seed-fern, is found fossilized in South America and southern Africa, as well as in Antarctica, India, and Australia. Through the study of past and present animal species numerous examples of organisms have been unearthed that are now separated by large expanses of water, but appear to have originated from the same place. For example, *Pliolophus*, the ancient predecessor to the horse, has been discovered in Upper Paleocene and Eocene rocks in North America and in Lower Eocene rocks in Europe,

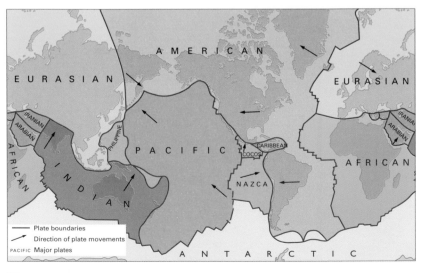

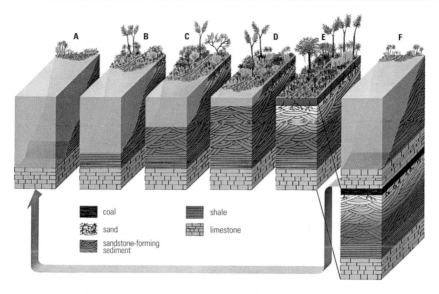

coal

sand

sandstone-forming
sediment

shale

limestone

suggesting that it migrated to Europe when the continents were joined, or were at least separated by a much smaller Atlantic. There is also evidence to suggest that the Caledonian mountain range in Scotland was once part of a continuous chain running through Europe and North America.

SEDIMENTARY ROCK TYPES

Fossils are usually associated with sedimentary rocks. Sedimentary rocks are formed through weathering and chemical processes whereby rocks and minerals have been broken or worn down into fragments and then compressed.

Some were formed, grain by grain, of preformed rock that was weathered, eroded, and deposited elsewhere, either by wind, rivers, or glaciers. Others settled as vegetable matter or dead organisms on river or sea beds. Others arise through chemical reaction.

The action of waves pounding on a beach will result in rocks composed of fragments of particles, but in this case the particles will be more rounded. Geologists are able to judge the age of a sandstone based on the degree of roundness of individual particles.

Rocks composed of rock, mineral, and fossil particles are termed **clastic rocks.** They are subdivided according to grain size. There are three

Sediment is laid down in a specific order that may be endlessly repeated if the region where deposition takes place is sinking. Limestone deposits cover the sea bed when a delta is too distant to be influential (A). As a delta encroaches (B), fine-grained muds that become shale are deposited, followed by coarser, sandstone-forming sediments (C) as the advance persists. As the water shallows, current bedding (D) indicates that sand is being deposited. Once the delta builds above water level (E) it supports swamp vegetation, which will eventually form coal. When the region sinks (F), the cycle restarts.

31

main groups. **Mudrocks** and **sandstones** usually form in deserts, rivers, and seas. **Conglomerates** are sediments with large rounded particles (some boulder-sized), while angular particle sediments are known as **breccias**.

Non-clastic rocks are those formed by the precipitation of minerals and/or accumulation of plant and animal remains. They include many **limestones** and the evaporites. They are subdivided on their mode of origin, chemical composition, and textures.

Black shale is often rich in fossils of mollusks and plankton.

Clastic rocks
Mudrocks
These are the finest grained of the clastic rocks, and individual grains are not visible to the naked eye. The main constituents are **clay minerals**, **quartz**, **feldspar mica**, and **calcite**. These are also described as **argillaceous**.

Clays are the very finest of mudrocks. Clay is turned to claystone by compaction, when water between the grains is squeezed out by the weight of more sediment accumulating above. Muds and clays often yield rich faunas of **mollusks**, small **corals**, and **brachiopods**.

Shales are compacted mudstones (a fine mudrock) in which the flat minerals, including clays and micas, are realigned along a preferred plane of orientation. This often gives the shale a degree of flakiness and a surface sheen or glitter.

Siltstones may contain quartz, feldspar mica, and calcite. They have a low clay content. **Brachiopods** and **bivalves** are among the more common fossils associated with these sediments.

Generally the finer the grain size, the weaker the current that carried the sediment, and the deeper the water in which it was deposited. The presence of **pyrite** also may suggest the absence of oxygen and bacteria. A low level of water movement and few agents of decay will therefore give rise to well-preserved fossil material. This is often true of fine-grained rocks, with **black shales** containing the remains of many organisms that floated or swam in open seas. However, few **seafloor** organisms could be expected to live under such inhospitable conditions. **Graptolites**, **goniatites**, and thin-shelled **bivalves** are common fossils in black shales of the Paleozoic era.

Sandstones

Unlike the finer grained mudrocks, the composition of various sandstones can be determined quite well with a hand-lens. Individual grains can vary from angular to well rounded, the degree of roundness indicating its age. Sandstones consist mainly of three main components: **quartz, feldspar,** and **rock fragments.** Quartz is a harder, more stable mineral than feldspar and will survive more than one or two cycles of weathering, erosion, and deposition.

Quartz sandstones

Sandstones consisting of well-rounded quartz grains are among the most mature sediments. Quartz sands are often porous and the fossils they contain may be dissolved to leave casts and molds. Vertical burrows, wood fragments, and shelly faunas are the most common fossils in such sandstones.

The color of sandstones can be quite distinctive, but color does not necessarily indicate environment. The color is related to the chemistry of the cement and the constitutional minerals. Red sandstones are often the result of the **oxidation** of iron. Sandstones are formed on land, in rivers and deltas, and in shallow seas. They often indicate formation in relatively high-energy conditions, often involving wind and water. Terrestrial sandstones may contain plant remains, whereas materials deposited from water will often retain abundant evidence of both **trace fossils** and **body fossils**.

Arkose

This descriptive term is used when detecting the presence of **feldspars, quartz,** and **rock fragments** within a sediment of sandstone grain size. Feldspars would account for 25 percent or more of the rock, and rock fragments 50 percent or less.

Lithic sandstones

These are sandstones in which **rock fragments** account for more than 50 percent of the rock. They also have a very low matrix percentage. Other sandstones can be classified on the presence of a specific mineral. **Glauconite, mica,** and **phosphate** can characterize a specific sandstone, and each indicates a particular environmental association. Muds and minerals such as **iron, calcium carbonate,** and **silica** can cement the grains within a sandstone.

Arkose *is the oldest fossil-bearing rock in Connecticut, US.*

33

The terms **argillaceous sandstone**, **ironstone**, **calcareous sandstone**, and **siliceous sandstone** are therefore used to describe such sediments.

Conglomerates and breccias

These rocks are coarse-grained. **Conglomerates** are recognized by the presence of surrounded to rounded grains, **breccias** by angular ones. Once again, the degree of rounding is an indication of maturity. Conglomerates can be classified by the type of pebbles within them. They are often linked with stream, lake, or seashore environments. Due to their angular fragments, breccias indicate a limited degree of transportation. Fossils are rare in both rocks. These coarse-grained rocks are also termed **rudaceous**.

Nonclastic rocks

This group includes a variety of rocks formed by the precipitation of minerals and plant remains. For paleontologists, the most important nonclastic rocks are **limestone** and **coral**.

Limestone

Limestones are rocks with a higher proportion of **calcium carbonate** than other constituents. **Calcite** is the most common mineral, which may occur as very fine crystals or in the skeletons and shells of invertebrate animals. The degree of fragmentation of organic remains is an indication of the energy level at the time of deposition: broken fragments suggest a higher level of energy than whole specimens. The organic fragments are known as **bioclasts**.

Shelly limestone *is largely composed of fossilized shells.*

Microfossils belonging to the **Foraminifera** are common components of both fine- and coarse-grained carbonate rocks. The tiny shells of planktonic organisms settle on the sea floor and accumulate to form an ooze. The best known of these is the famous white chalk of northern Europe.

Shells of the larger Foraminifera are also major contributors to the formation of limestones. Different families dominate specific episodes of geological time. For example, the many chambered shells of the coinlike **Nummulites** are characteristic of **Tertiary** Mediterranean limestones. The larger Foraminifera are mostly indicators of shallower water environments. Accumulations of **gastropods**, **bivalves**, and **brachiopods** are common in the **fossil record**. The term 'shelly **limestone'** describes limestones rich in such fossils or their debris. Accumulations of shells or shell debris are also referred to as **'coquinas'** or '**lumachelles**.' Brachiopods are common in limestones of the Paleozoic era, whereas bivalves and gastropods are more important in limestone of the Mesozoic and Cenozoic eras. Worms and **echinoderms** are also involved in the accumulation of carbonate sediments.

Coral

Corals are major constituents of limestones during the Paleozoic, Mesozoic, and Cenozoic. They are often associated with algae and **bryozoans** as reef-builders, with many other organisms encrusting or boring into the fabric of the build-up. Reeflike mounds may also be formed by **rudist bivalves** and **archaeocyathids**.

Algae have been important 'rock-formers' for billions of years. **Stromatolites** – layered, domed structures common in the Precambrian and Early Paleozoic periods – were formed when algae trapped grains on their sticky surfaces. In shallow waters and on beaches, the grains may be arranged concentrically or form asymmetric structures called **oncolites**.

Evaporites/phosphate rocks and nodules

Evaporites are mainly chemical rocks formed when dissolved salts, concentrated by water evaporation, precipitate out as massive or **nodular** deposits. **Rock salt** or **halite** is a typical evaporite formed from the evaporation of saltwater lakes, lagoons, or shallow seas. **Gypsum** often grows as nodules in mudrock sequences, and is associated with **sabkha** (supratidal) environments. **Algae** and **gastropods** are the most common organisms associated with these types of deposit.

Most **phosphate** rock forms on areas bordering the edge of the continental shelf. The rock is often nodular or pelleted, with organic materials often replaced by phosphate materials. The areas of deposition were rich in microscopic plant materials and, as a result, fish and large fish-eating animals are common. Phosphate rocks are frequently rich in **vertebrate** remains and **trace fossils**.

Phosphate rocks yield many marine fossils.

EVOLUTION

Planet Earth coalesced from a cosmic cloud about 4,600 million years ago, but it took hundreds of millions of years for conditions to stabilize enough for organic molecules to accumulate. Earth's early atmosphere was devoid of oxygen, made up mostly of hydrogen, ammonia, methane, and water vapor. This thin layer was no shield against the Sun's powerful radiation. Lightning storms, volcanic eruptions, and meteorites were commonplace, and all provided energy vital to the evolution of life.

Evolution can be defined as the changes that occur across successive generations of organisms. The causes of evolution include **natural selection** and **genetic drift**. Early work on evolutionary theory was initiated by **Jean Lamarck** during the early 1800s, but it was not until the mid-1800s that the theory was considered worthy of greater attention. Quite independently, **Charles Darwin** and **Alfred Wallace** developed the same theory of evolution. They observed that organisms produce far more offspring than they need to maintain the size of their population. Yet most populations remain relatively constant in numbers because many die due to predation, disease, and starvation. Consequently, individuals are competing with each other to be the one to survive. Each individual has different **genes** and is, therefore, distinct from the others. Some individuals will be better suited to survive in the existing conditions – a situation known as '**survival of the fittest**.' These 'fitter' individuals are more likely to breed and pass their advantageous genes on to their offspring. Over many generations, the individuals with favorable characteristics will build up in number at the expense of those lacking them. In time, more **variations** will lead to the evolution of a new species. This evolutionary process is known as 'natural selection.'

Evidence for the theory of natural selection is that dated fossil remains show that life did not arise at once, but as a gradual change from one type of organism into another. Furthermore, the structures of different animals or plants show such similarity that it is highly probable they evolved from a common ancestor. For example, The bones in the wing of a bird, the arm of a primate, and the paddle of a whale, all show remarkable similarities. Equally, many of the proteins in organisms are fundamentally the same, and we all share many common genes.

Species are not evenly distributed around the world. Elephants are found in parts of Asia and Africa, but not in similar habitats in South America. This discontinuous distribution is explained by the theory of evolution. A species originates in a particular area and individuals disperse to avoid overcrowding. As they meet new environments, they adapt to the new conditions, but climatic, physical, and other barriers prevent them from breeding with their ancestors. Thus a new species is created, which continues to adapt to the new conditions.

THE FOSSIL RECORD

The world is constantly populated by potential fossils. After death, however, only a small percentage of living organisms will be preserved. Soft-bodied creatures will mostly vanish without trace, and even animals with hard parts will usually be destroyed or fragmented. The material preserved will, therefore, provide an unbalanced representation of life as we know it today. The same is true of the fossil record. It is incomplete, and many families have undoubtedly vanished without trace.

PRIMORDIAL SOUP – HOW LIFE BEGAN

All living things are composed of carbon-based organic molecules and, crucially, are capable of reproducing themselves. These characteristics, which typify life, first developed in simple molecular systems some 600 million years after the Earth's formation.

Some clues as to how these molecules were first formed on the early Earth come from laboratory experiments. But only a few of the simpler building blocks of life – such as the **amino acids** that make up **long protein molecules** – have to date been produced in such experiments. These simple organic compounds accumulated in the ancient seas and, warmed by the Sun, this 'prebiotic soup' formed the larger and more complicated molecules – for example, **nucleic acids**, **proteins**, **lipids**, and **polysaccharides** – that make up living cells. Larger organic compounds formed molecular systems, capable of storing information about their structure in a way that identical systems could be reproduced. Just what such self-propagating systems were like is unclear because this primary stage of life has left no fossils behind in the rocks.

The first living cells probably arose about 3,500 million years ago and may well have been the result of spontaneous molecular aggregations. Protenoid microspheres, shown here, are small, spherical aggregates of protein that can be artificially made by heating amino acids. These spheres have certain properties of cells, suggesting that aggregates like these may have been involved in that first step of life on Earth.

SIMPLE CELLS

The first 'cells' might have been formed when hollow spheres of self-sealing fatty membrane coalesced around groups of self-replicating molecules.

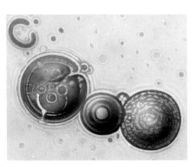

For almost 2000 million years, simple **unicellular microorganisms** were the only forms of life on the planet. Their remains are sometimes found in fossil **stromolites,** structures laid down from successive layers of cells and trapped debris. Some of these early cells developed the ability to **photosynthesize**, giving out **oxygen** as a waste product, and in time producing an atmosphere rich in oxygen.

The next milestone in evolutionary history was the appearance of much more highly developed cells – **eukaryotic cells** – around 1,500 million years ago. From these cells, which have a nucleus and complex internal structure, evolved single-celled **protozoa** and **algae**, and all multicellular life.

THE CAMBRIAN EXPLOSION

The earliest traces of multicellular animals are rare imprints of soft-bodied invertebrate animals in rocks from around 600 million years ago, toward the end of the **Precambrian** period. They resemble jellyfish, segmented worms, and sea pens. Some scientists think they may represent an entirely different form of

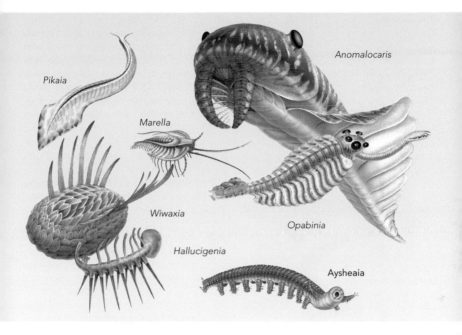

Pikaia

Anomalocaris

Marella

Wiwaxia

Opabinia

Hallucigenia

Aysheaia

body organization corresponding to a failed evolutionary experiment.

In general, only the hard parts of an organism – shells, scales, spicules and, later, bones – become fossilized. So the **fossil record** is incomplete and highly selective, incorporating only very rare traces of the many entirely soft-bodied animals and plants that must have existed. Invertebrate animals with hard parts started to appear at the beginning of the Cambrian, the period that saw an explosion of animal life in the oceans.

HOW ANIMALS EVOLVED TO LIVE ON THE LAND

About 520 million years ago, at the beginning of the **Cambrian** period, representatives of most of the main groups (phyla) of animals had appeared. This explosion of animal life, however, remained confined to water until past the end of the Cambrian, some 500 million years ago. The first forms of life on the land were mats of **algae**, **lichens**, and **bacteria**. They managed only to colonize the edges of shallow

The Burgess Shale, western Canada, provides a fascinating glimpse of marine life 570 million years ago. These creatures lived during the Cambrian explosion, a period of intense evolutionary diversification when the ancestors of probably all the modern animal groups we know today came about.

39

The evolution of the pentadactyl (five-fingered) limb can be partly traced through the fossil record. The basic design was already present in lobe-finned fish. Adapting to land, the limb rotated downwards and away from the body.

pools. But around 400 to 500 million years ago they were followed by the first true land plants. This simple vegetation was in turn colonized by the first known wave of air-breathing land animals: tiny millipede-like **arthropods**, protected from drying out by their hardened outer skeletons. The first animals with backbones – the **fish** – evolved in the oceans of the **Late Cambrian**. One line of fish with a bony skeleton developed air-breathing lungs and 'limbs' strong enough to support them on land. They gave rise to the first four-legged vertebrates – the **amphibians** – from which all future vertebrate animals evolved.

The first amphibians to emerge from their freshwater habitat found low-lying, swampy, open forests of tree-like horsetails and club mosses, liverworts, and other small plants. **Reptiles** evolved from one of the amphibian groups and were able to make much better use of

Eusthenopteron

1 Drepanaspis (jawless fish)
2 Platysomus (ray-finned fish)
3 Eusthenopteron (transition from fish to amphibian)
4 Icthyostega (early amphibian)
5 Diadectes (early amphibian)
6 Meganeura (prehistoric insect)
7 Pareiasaurus (early reptile)
8 Icarosaurus (gliding reptile)
9 Thrinaxodon (transition from reptile to mammal)
10 Archaeopteryx (gliding reptile)
11 Tyrannosaurus (largest carnivorous dinosaur)

Ichthyostega
the limb is compact and close to the ground

Seymoura
extended limbs for greater articulation and ground clearance

The Triassic saw the beginning of the 'Age of the Dinosaurs,' and for more than 150 million years, no other animal larger than a hen walked the Earth. Other reptile groups evolved at this time, one of which evolved into warm-blooded mammals. As dinosaurs grew larger, some, such as Kentrurosaurus, evolved spiked protection against predators; others, including Compsognathus, relied on speed and developed a bipedal gait. Archaeopteryx was an early feathered bird-like creature from the Jurassic. It shared the sky with the Pterodactylus. Around the beginning of the Cretaceous, flowering plants evolved.

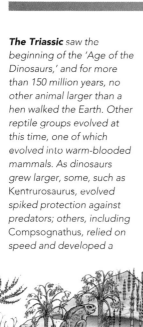

By the Devonian period, vertebrate fishes had evolved a number of separate groups. Extinct placoderms (platedfish), such as Dunkleosteus, swam alongside ray-finned fish, such as Platysomus. Lobe-finned fish, for example Eusthenopteron, made the transition to air-breathing amphibians like Ichthyostega. The lush vegetation of the Carboniferous swamps supported a profusion of animal life. Amphibians exploited the environment, and some became increasingly terrestrial. One such group evolved the trick of reproducing away from water by means of tough-shelled eggs, and these became ancestral reptiles (denoted by the suffix 'saurus.') Animals increased in size, with large carnivores such as Diadectes feeding on herbivores such as Pareiasaurus. Insects took to the air and gave rise to air-borne predators like Meganeura and Icarosaurus.

41

the land, filling every available habitat and ecological niche. They became adapted to many different ways of life, taking to the air as **pterodactyls**, and even returning to rule the water for a time, as did the **plesiosaurs**, **ichthyosaurs**, and other forms.

The **Mesozoic** era, stretching from the end of the Permian period (250 million years ago) to the end of the Cretaceous (65 million years ago), is often called the '**Age of Reptiles**.' The earliest mammals also appeared, even as the reptilian **dinosaurs** were rising to prominence, but they remained small and inconspicuous for millions of years.

Pterodactyls had flimsy, membranous wings. It is believed that they were gliders, incapable of sustained flight.

EVOLVING TO EXTINCTION

Throughout evolutionary history, many new species have appeared, and many plants and animals have also disappeared, becoming extinct, so that the many millions of present-day species represent only a small part of all the living things that have ever existed. Extinction, however, is not always gradual. The history of life has been punctuated by several periods of **mass extinctions**, when large numbers of species become extinct over a very short time – in geological terms. One such mass extinction occurred at the end of the **Permian**, when up to 96% of marine species are estimated to have become extinct, including the **trilobites**. But the best known example of extinction happened at the end of the **Cretaceous** period, when the **dinosaurs** and many other species vanish from the **fossil record**.

Plesiosaurs were a successful group of marine reptiles that became extinct at the end of the Cretaceous.

Kronosaurus was a short-necked Plesiosaur. It lived in the seas that covered parts of Australia during the early Cretaceous.

The most likely explanation for the demise of the dinosaurs is that Earth's climate went through a major change and became much cooler. It has

► **Brachiosaurus** was one of the biggest land animals ever.

▼ **Iguanodon** was a plant-eating dinosaur from the early Cretaceous.

► **Plateosaurus** was the first of the great long-necked herbivores.

◄ **Lambeosaurus** belonged to a group of 'duck-billed' dinosaurs common in the Upper Cretaceous.

▼ **Stegosaurus** had 17 bony plates running down its back and tail. These may have been used in mating displays or to regulate body temperature.

▲ **Pachycephalosaurus** was a dome-headed, plant-eating dinosaur alive during the Upper Cretaceous.

▲ **Ankylosaurus** was an armoured dinosaur with a thick tail-club used to defend against carniverous dinosaurs.

► **Triceratops** was a rhinoceros-like dinosaur with three head-horns and a large bony plate (frill) on the back of its head.

43

been suggested that this was due to the impact of a gigantic meteorite, which threw up clouds of dust into the atmosphere. Dinosaurs lacked the sophisticated system that mammals possess to maintain a warm body temperature, and probably could not stand the dramatic change.

HOW MAMMALS AND BIRDS EVOLVED

Extinct elephants and the modern elephant differ from each other in the shape of their heads, as shown by the skulls of the primitive Moeritherium, *the* Trilophodon, *and the modern elephant. The* Trilophodon *had four tusks, two in the upper jaw and two in the lower.*

From a present-day point of view, the world and its flora and fauna would have begun to look increasingly familiar after the end of the **Cretaceous** period (65 million years ago). Flowering plants and trees flourished, insects had diversified into their modern forms, birds flew in the air, and small mammals walked, scurried, ran, and hopped over the land. During the succeeding **Tertiary** sub-era (up to 2 million years ago), rainforests, temperate broad-leaved forests, and, later, expansive grasslands provided new habitats into whichthese novel creatures could spread.

But mammals did not have the land entirely to themselves. Predators believed to have preyed on small mammals included giant flightless birds, such

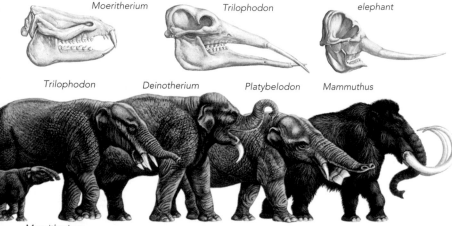

Moeritherium   Trilophodon   elephant

Trilophodon   Deinotherium   Platybelodon   Mammuthus

Moeritherium

The oldest ancestor *of the modern elephant was the small, tapir-like* Moeritherium *of the Eocene epoch. The* Trilophodon *comes from the*

Miocene *epoch, as does the* Deinotherium, *which had a pair of tusks attached to the lower jaw. The* Platybelodon, *also from the Miocene, had*

flattened lower tusks shaped like a shovel. The woolly mammoth Mammuthus *was adapted to the cold of the Pleistocene epoch.*

as **Diatryma** (during the early **Eocene** period) and the South American **Phorusrhacos** (in the **Miocene**).

During the Tertiary, mammals spread all over the world and evolved into many different types. Large, fleet-footed, hoofed mammals roamed open grassy plains, preyed upon by swift carnivores. Bats took to the air. The ancestors of dolphins and whales returned to the oceans from which their remote ancestors had emerged several hundred million years before. The early primates took to the trees, where their precarious lifestyle led to the evolution of sharp stereoscopic vision, delicate control of hands and feet, and an enlarged brain. From their descendants evolved the line leading to the great apes and to human beings.

Archaeopteryx is the earliest known recognizable bird. It dates from the Jurassic period. It is, in fact, a reptile and the wings were elongated forelimbs, complete with claws.

MONOTREMES AND MARSUPIALS

The very first mammals probably laid eggs, like their reptile ancestors and like the primitive egg-laying monotremes that survive today – the platypus and echidna (spiny anteater). Different species of echidna are also found in New Guinea. These early mammals gave rise to the marsupials, which once lived mostly in South America, where 70 species still exist. Today, Australia probably supports the greatest number of marsupials. On this island continent, which became isolated by **continental drift** before the later placental carnivores could reach it, marsupials were able to evolve further without competition.

As mammalian orders evolved, many species progressively grew larger in size. Good fossil records for horses, for example, show how they developed from relatively small animals, most unlike their modern forms. In some orders of mammals, elephants, and rhinoceroses for example, giant forms developed that have since become extinct. Many other groups of animals also had members much larger than present-day examples.

Platypus are sometimes called 'living fossils' because of their ancient heritage and unique appearance. The earliest fossil relative of the platypus has been found in rock more than 60 million years old.

45

Fossil hominid skeleton of Australopithecus afarensis, *popularly known as 'Lucy,' discovered at Hadar, Ethiopia, in 1974. She dates from 3.3 million years ago, and is widely accepted as the earliest link in the human record. The remains comprise 40% of the entire skeleton and include skull fragments, a mandible, most of the left and right arm, vertebrae, rib fragments, sacrum, left pelvic bone, left thigh, and right lower leg. The form of the pelvic bones showed her to be female. Erupted wisdom teeth suggested she was 20 years old, and the thigh bone indicated she was small, 4 feet (between 107 and 122 cm) tall.*

A   B

THE EVOLUTION OF MAN

The fossil record of human evolution is patchy and unclear. Some scientists believe that our ancestry can be traced back to one or more species of *Australopithecines*, which flourished in south and east Africa between four and one million years ago. Other scientists believe that we are descended from some as yet undiscovered ancestor. The earliest fossils that can be identified as human are those of *Homo habilis* (handy people), which date from two million years ago. The earliest fossils of our own species, *Homo sapiens* (wise people), date from c.250,000 years ago. An apparent side-branch **Neanderthals** (*Homo sapiens neanderthalensis*), existed in Europe and west Asia c.130,000–30,000 years ago. Fully modern humans, *Homo sapiens sapiens*, first appeared c.100,000 years ago. All human species apart from *Homo sapiens sapiens* are now extinct.

Although the fossil record is not complete, we know that humans evolved from ape-like creatures. Our earliest ancestor, Australopithecus afarensis [A], *lived in north-east Africa some 5 million years ago. Over the next 3–4 million years, Australopithecus*

africanus [B] evolved. Homo habilis [C], *who used primitive stone tools, appeared c.500,000 years later. Homo erectus [D] is believed to have spread from Africa to regions all over the world 750,000 years ago. Records indicate that from Homo erectus*

evolved two species, Neanderthal man [E], who died out 40,000 years ago, and could have been made extinct by the other species, the earliest modern man Homo sapiens sapiens [F].

FOSSIL COLLECTING

In order to get the most out of your fieldwork, it is important to plan ahead. One of the first items to acquire is an **accurate map** of the area. For the best results, consult scientific journals for descriptions of the area, the outcrops, and the fossils found there. They will often contain grid references, and pinpoint the locality with great accuracy. Should your trip take you to the coast, consult **tide tables**.

WHAT YOU WILL NEED

Strong bag
Geology hammer and chisel
Eye protection
Gloves
Plastic bags
Tissue paper/ paper towels and newspaper
Strong tape
Notepad and pencil
Suitable clothing: i.e. boots and waterproofs, hat
Compass-clinometer
Camera
Map

FINDING FOSSIL-RICH BEDS

Fossil collection is recommended **only from fallen material**, and the best place for this is along the shore line at a beach, especially after a storm that may have caused new material to wash up. Collecting from cliff faces or any large expanse of rock can be both damaging to the natural environment and dangerous. Before you make an extraction, make sure that it is really necessary. Large accumulations of bedded fossils may be better off where they are, in which case a photograph and accompanying notes will be sufficient information for your collection. The use of hammers is largely seen as unnecessary and potentially damaging. Quarries, rocky outcrops, and coastlines are usually clearly defined on both geological and topographical maps. Many geological maps also indicate whether or not the rocks in these areas are fossiliferous.

Once you have decided on the general area you are going to visit in search of fossils, you need to know

Field geologists carry in their backpacks a notebook, pens, pencils, map, compass, and camera. Other less essential items include a hand-lens, brush, sieve, penknife, plastic bags, safety googles and plastic tubes. By and large, the use of a hammer and chisel is unnecessary.

exactly where to begin looking for material. This can be a problem in some areas, but usually there are clues to help you. Often you will notice that the rocks are layered and dip in certain directions, or they are horizontal. Differences in the color and texture of the rock may tell you that different beds alternate with each other – the boundaries between them representing the **bedding planes**. If exposed as a flat surface, a fossil-rich bedding plane will invariably yield the best materials and data. Specimens will be etched out over the surface, and in some cases will be easy to collect.

If the bedding planes are difficult to reach, then the presence of fossil material may still be observed along fractured or jointed surfaces. It is usually difficult to collect fossils exposed in this way, because they are often embedded in protected, unweathered surfaces. You can, however, use this evidence to trace the bed until a weathered surface or bedding plane is exposed. You may also use features (such as color, type of weathering, and bed thickness) to trace a **fossiliferous horizon** along the strike of the bed or between outcrops. In limestone areas, fracturing along joints, due to ice, may result in an accumulation of fossiliferous sediments on a scree slope. This is true for other well-cemented and well-jointed sediments that occur as outcrops in mountainous areas.

On the seashore, soft rocks will often collapse and flow, and finding a bedding plane is difficult. Often the fossils are buried in very thick sediments, and discovery is mostly by chance. At some localities, it is better to search along the foreshore, since the fossils washed out of the cliff face tend to accumulate near the high-tide mark. **Strandlines** are often excellent sites for fossil collecting. It is also possible that a fossiliferous horizon will be exposed on the beach, and a search over the flat rock pavement, which may extend outwards along the beach, will prove more rewarding than using a hammer and chisel on a vertical cliff exposure. Even in areas where the rocks are vertical and the outcrop washed and rounded, a **rock grain** will be apparent in most sediments. Look for limitations, as well as color banding and the other features. These will often be related to bedding, and careful probing will reveal well-preserved material. If you cannot find the bedding plane and the rock is not well jointed, it may be a waste of time trying to collect from that particular outcrop, but remember to record what you can anyway.

RECORDING DETAILS

A compass-clinometer measures strike and dip. Always try to place a notepad along the bedding plane to be measured, so any irregularities in the surface are evened out. Place the edge of the compass against the notepad, and rotate the compass until the bull's-eye level is centered. Read the compass at the white end of the needle (which always points north), noting whether an imaginary center line through the compass lies to the east or west of the white tip. For example, if the center line lies 20° west, then record the strike as "N20W." Now use the clinometer to measure the angle and direction of the dip. Dip is always perpendicular to the strike: a strike of N20W could only dip to the NE or SW, never to the SE or NW. If it dips 20° south, record N20W/20S.

First, stand back and observe the exposure as a complete entity. Then draw a sketch and describe briefly the major features. If the rocks are well bedded or show signs of cross-stratification, then measurement of the dip and strike should be recorded. The **dip** is the amount by which the bedding plane is inclined to the horizontal. The **strike** is drawn at right angles to the dip. (A simple analogy is a roof: the ridge representing the strike, and the roof-slope the dip angle). One method often used in the field is to pour water gently onto the inclined surface: the line it follows will normally represent the dip direction. In order to take bearings of your location you should sight the compass across to the first of several prominent landmarks, align the compass to north, and take a reading. Repeat this exercise for subsequent readings from other landmarks approximately 90 degrees away from the first one. Other field workers will then be able to use your bearings and, by adding 180 degrees to each, draw lines from the landmarks to locate the position with accuracy.

Once you have recorded your position accurately and taken measurements, study the relationships of the individual beds, or layers of strata.

They may be described in terms of their thickness and the contact they make with one another. Thin, medium, and massively bedded are terms used in the description of a bedded sequence. Sketches will again help record the interesting features you observe. Once you have completed this task, you can then describe the rock and fossils it contains. Remember to use descriptive terms in a scientific manner: for instance, "fine grained, buff-colored limestone, which contains rounded rock fragments and rare, small bivalves."

Once you have completed these important tasks, other items related directly to fossil collection remain. For example, if the rocks are rich in fossils, sketches and photographs of their position and relationships

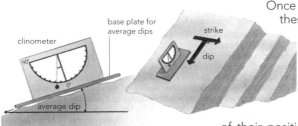

base plate for average dips

clinometer

strike

dip

90°

0°

average dip

should be made. Accurate measurements of the size and shape of individual fossils should also be entered into your notebook. Note also the position of the fossil. Is it loose or in situ? Has the organism been preserved in life position or transported? The use of a hand-lens is essential for spotting fine details. **Faunal communities** can be recorded in terms of the number and position of species. Count and measure the species within a one yard (one-meter) square placed at regular intervals along a given line on the bedding plane. This data will be invaluable to others, and will provide you with an in-depth knowledge of the specific horizon and of the time-interval studied.

EXTRACTING MATERIAL

The destruction of an outcrop in search of fossils is irresponsible (see Fossil Codes, page 186). Fossils should be collected sparingly, preferably without hammering. Often the best fossils are those that have been weathered out over a long period. These rest on the rock surface or among the scree at the foot of an outcrop. Collecting is a matter of judgement. If you think that the removal of the specimen is essential then do it, but first consider its scientific value and whether or not it would otherwise be damaged or destroyed by the elements or by vandalism.

Extracting a specimen embedded in a rock may prove to be a long and difficult process. Indeed, on occasion even the most beautiful specimen will have to be left behind since its collection would create too many problems, or even become hazardous. Before you attempt to collect a specimen make sure that it is feasible to remove it, and that the use of a hammer will not destroy either the specimen or the outcrop. Your approach to the collection of a specimen will vary in relation to the rock type and the nature of the fossil. In hard, well-jointed rocks the best approach is to exploit the surrounding joints. Usually the fossils collected from such well- cemented rocks will not need any further preparation in the field. If it is impossible to exploit joints when extracting a specimen, make sure that you chisel away from the specimen, leaving a generous margin of sediment around it before trying to cut underneath. Be sure to leave sufficient rock beneath the specimen to protect against fracture.

Faunal community of ammonites in a sample of rock from Marston Magna, Somerset, England. Promicroceras martonense lived in the Lower Jurassic (206–172 million years ago).

51

Specimens collected from soft sediments may be delicately preserved. It is essential, therefore, that they are lifted in a block of sediment, so that their support is not removed until you get home. To do this it may be better to use a sharp knife rather than a hammer and chisel. Simply carve away the sediment. Remove the material at the side and back before you attempt to cut away the underlying sediment. When free, place the specimen in a box and support it with soft tissue paper.

When collecting from soft sands or sandy clays, it may be better to **sieve** rather than scrape or dig into the material. In this way even the smallest specimens can be recovered and you will thus obtain a more balanced representation of the fossils preserved at that particular locality. Wet sieving is a technique frequently used for the collection of small mammalian fossils. Small fossils lying freely can be picked out with fine tweezers.

After collecting your samples, **number** them immediately, and record the number and **location** in your notebook. Make sure to wrap your material carefully in tissue and newspaper. Never throw them all unnumbered into a box. Tissue paper, newspaper, and strong tape are essential items for the paleontologist in the field. Other items include a variety of small brushes for the removal of loose sediments and some **polyvinyl-acetate (pva)** glue for application to weakened specimens. Large specimens may need to be surrounded by a papier-mâché/plaster of Paris jacket before removal.

BACK HOME

There are many techniques you can use to clean, prepare, preserve and enhance your fossil material. A good wash with water and a brush will often clean the more robust samples, but others should be treated with great care. Just brush these and pick away any loose sediment with a strong pin. A coat of thin polyvinyl-acetate (pva) glue will protect the exposed surface.

Some laboratories are well equipped with air abrasive tools, air dents, and dentists' drills. With these it is possible to take away surplus sediment easily, but once again care is needed, especially when working close to the specimen.

Skilled preparateurs use a combination of these tools

and diluted acids. The latter can be extremely dangerous in inexperienced hands, and care is needed when using these chemicals at home. **Limestones** respond well to treatment with **dilute hydrochloric** or **acetic acids** (two to 10 percent). **Silicified** fossils and bones can also be prepared from their sediments by using these chemicals. The specimens should be dry and the exposed parts coated with pva glue. Do not leave the specimen in acid for too long – just four to six hours at a time. When you remove it, wash it by leaving it in a bath of clean water for an equal period of time and then dry it completely. **Broken** or fractured fossils can be mended, but limestone fossils should not be fixed with an adhesive containing ascetic acid as this corrodes calcareous fossils. Do not wash fossils from unconsolidated **argillaceous shales**. Instead use alcohol and a soft brush to avoid turning the specimen into shapeless clay. The exposed areas of the fossil should again be coated before the process is continued.

CURATION AND AFTER CARE

When the preparation of your fossils is complete, it is worth taking the time to curate them properly. Make sure the number you placed on the specimen is transferred, together with the location details, to your **collection book** and **specimen cards**. A card should exist for each fossil. Apart from the location details, it should be marked with the drawer and tray numbers in which the specimen is kept. Other information, such as the **name** of the specimen, the **family** to which it belongs, and the **stratigraphic age** should also be recorded. Establish a cross-index system and your collection will be a credit to you and science. Over the years, you will be able to take pride in it and compare your material with that of the major museums. Other paleontologists may wish to visit you and use your material. On the other hand, specimens lacking all the details noted above will be of little or limited value. Update your information at regular intervals: names may change and outcrops disappear. If they do, record the details. Check also that the specimens are kept dry, and that **lyricized** material does not decay. This is difficult to prevent, but soaking in a **bacteriocidal disinfectant** may help. Remember to dry the specimen concerned and coat it again with pva glue.

Stages in preparing a body fossil
1. remove loose rock
2. immerse in dilute hydrocholoric acid (two to 10%)
3. Wash in deionized water
4. Paint with pva glue

CLASSIFICATION

The sorting of organisms of similar appearance is called **classification**. **Taxonomy** is the science (the theory and practice) of naming and classifying organisms.

Unlike the vast majority of living plants and animals, fossils do not have popular names. Their names are derived from Latin or Greek, and may be regarded as scientific names. Popular names such as 'dinosaur,' 'clam,' or 'ammonite' define groups, but not individual organisms. Scientific names are precise, and are used to describe the characteristics of the fossil and define the group to which it belongs. Every animal or plant, however large or small, belongs to a **species** (in other words similar organisms that can interbreed to produce fertile offspring). Several species may belong to one **genus**. These will be similar in overall shape or character and are closely related. The scientific name of the organism consists of two parts – a generic name and a trivial or specific name. Modern man is therefore named *Homo sapiens* in accordance with this procedure. **Genera** are grouped into **families**, families into **orders**, orders into **classes**, and classes into **phyla**. Each level of classification means a broadening of the characteristics that associate the constituent groups or **taxa**. When you label your specimens, it is essential that you give them the correct name. For most specimens this will already exist and can be found in books such as the *Treatise on Invertebrate Palaeontology*. Should you have difficulty in finding the right name, then seek help from your local museum or college. It may be that you have found something unique. If so, it is necessary that you follow the correct procedure and describe your specimen in a recognized scientific journal.

WHAT IS A SPECIES?

The most common definition of a species is a group of individuals that have the potential to interbreed freely to produce fertile offspring. Conversely, true species should be unable to breed successfully with members of other species. In a very few animal species and more commonly in some plant species, boundaries seem to break down, allowing successful breeding between members of different, closely related species. These are exceptions, however, and successful breeding is not possible between members of the majority of different species. In the case of fossils it is obviously not possible to test for interbreeding, and even in many living organisms, interbreeding potential is not known, or not easily tested. In practice, the most common approach therefore is to define a species on the detailed similarity of form and anatomy, and sometimes also lifestyle and behavior, shown by a group of individuals occurring in the same place at the same time (a natural population). For fossils, this place should be a single geological layer within a single small study area, with evidence that the fossils were once a **living assemblage**, rather than, say, an aggregation of shells brought together from many different places by currents and waves.

IDENTIFICATION KEYS

The recognition of organisms from fossil remains can be helped considerably by using a system of keys such as we provide below. Remember, however, that your specimen may consist only of a fragment of the complete animal, and so you must try and visualize the creature in its entirety. The first part of the key enables identification of a particular phylum or class to be made. Then, by turning to that phylum or class in the Identification Guide, you can identify your fossil further by using a second key. Many types may be checked against the pictures in the Identification Guide.

Echinoderm
(crinoid)

Key to major fossil invertebrate groups

1a Solitary   ▶ **2**
  b Colonial   ▶ **3**

2a Non-chambered   ▶ **4**
  b Chambered   ▶ **5**

3a With pores   ▶ **18**
  b Without pores   ▶ **19**

4a Coiled   ▶ **GASTROPODS**
          (page 94)
  b Non-coiled   ▶ **6**

5a Coiled   ▶ **7**
  b Non-coiled   ▶ **8**

Coral

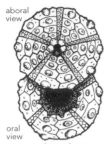

aboral view

oral view

Echinoderm
(echinoid test)

Mollusk
(ammonite *top*, belemnite guard *bottom*)

Trilobite
(dorsal view)

Sponge

6a Shell composed of single skeletal structure ▶ **10**
  b Shell composed of more than one skeletal component ▶ **13**

7a Chamber partitions straight or slightly curved ▶ **8**
  b Chamber partitions folded *see* **ammonites** (page 107)

8a Chamber partitions with small tubelike structure (the siphuncle) ▶ **9**
  b Small to microscopic organisms; siphuncle absent ▶**FORAMINIFERANS** (page 58)

9a Chamber partitions (septa) straight or slightly curved *see* **nautiluses** (page 106)
  b Chamber partitions slightly curved; major component of skeleton a bullet-shaped guard *see* **belemnites** (page 116)
  c Chamber partitions folded *see* **ammonites** (page 107)

10a Fossil characterized by radial symmetry ▶ **11**
  b Fossil not characterized by radial symmetry ▶ **12**

11a Vertical radiating partitions ▶ **CORALS** (page 62)
  b Solid with central tube radiate patterning *see* **crinoidea** (page 140)

12a Fossil large with large aperture; pores present; single wall ▶ **SPONGES** (page 59)
  b Fossil large with large aperture; pores present; double wall ▶ **archaeocyathids** *not illustrated*
  c Fossil large with large aperture *see* **gastropods** (page 94)
  d Fossil microscopic to small ▶ **FORAMINIFERANS** (Page 58)

13a Bilateral symmetry ▶ **14**
  b Radial symmetry ▶ **16**

14a Shell composed of two valves ▶ **15**

b segmented skeleton

▶ **ARTHROPODS**
(page 128)

15a Two valves, usually mirror images of each other and usually of equal size
b Valves usually of different size, not mirror images, equilateral symmetry

see **bivalves**
(page 118)

▶ **BRACHIOPODS**
(page 82)

16a Radial symmetry follows five-fold plan
b Radial symmetry without five-fold plan

▶ **17**

▶ **CORALS**
(page 62)

17a Skeleton with arms and without stem
b Skeleton rounded or plate-like; arms incorporated into test
c Flattened, five radiating arms

see **crinoids** (page 140) or **blastoids** (page 143)
see **echinoids**
(page 146)

see **asteroidea**
(page 144)

18 With single porous wall

▶ **SPONGES** (page 59)

19a Laminate structure; boxlike units
b Tube or boxlike units

see **stromatoporoids**
not illustrated
▶ **CORALS** (page 62)

20a Large with vertical radial partitions
b Small to microscopic; no radiating vertical partitions

▶ **CORALS** (page 52)

▶ **21**

21a Possessing rodlike branches
b Mosslike with many small apertures

▶ **GRAPTOLITES**
(page 150)
▶ **BRYOZOANS**
(page 75)

Brachiopod
(shell partly cut away at top)

Mollusk
(gastropod, shell partly cut away)

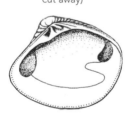

Mollusk
(bivalve, one valve removed)

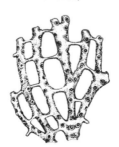

Bryozoan

The age ranges given here can be compared with the geological column on page 16. In this book, "Recent" is used for living organisms, together with their immediate fossil record, going back to the last 10,000 years. The geographical range is given with abbreviations as follows: NA – North America; SA – South America; E – Europe; Af – Africa; Aust – Australasia. Asia is not abbreviated. "Worldwide" indicates probable occurrence in all these regions. Where a geological age range is uncertain, a question mark is used.

FORAMINIFERANS

(Phylum *Protozoa*)

Foraminiferans are microscopic to unicellular animals. The majority of these single-celled creatures live in the sea. They frequently possess a skeleton or **test** that may consist of an organic, horny substance called **chitin**, of calcium carbonate, or of sand grains cemented together around the animal. The test varies in size from 1/50 inch to four inches (0.05 millimeters to 10 centimeters). Specific families have a characteristic test wall structure. Shape and size are good indicators of mode of life, and the number and arrangement of chambers help in the classification of genera and species. You will need a powerful microscope to find and study smaller foraminiferans. Larger forms exist throughout the stratigraphic record. Some of these are important rock formers and their abundance makes them easier to find.
Range: Cambrian to Recent

Nummulites *Paleocene: E Carib*

These larger foraminiferans are essentially disk-shaped and multichambered. The test is composed of calcium carbonate and perforated with small holes. The outer surface may be marked with a curved or spiral ornamentation and sometimes with small nodules. Nummulites are first found in Paleocene rocks and are most common in areas bordering the Mediterranean and Caribbean seas; they become extinct in the Oligocene.

Nummulites

SPONGES

(Phylum *Porifera*)

Sponges are many-celled organisms that rank just above the **protozoans** in order of classification. Structure usually radial with a central cloaca and surfaces covered with pores. The majority are marine animals that range in size from less than 1/2 inch to 3.2 feet (one centimetert one meter) in diameter. Although they have many cells, the sponges have no recognized tissue layers. The soft body is often supported by thin, rod-like elements called **spicules**. These may be separate or fused, and composed of either **calcium carbonate**, **silica**, or a horny substance called **spongin**. Individual, infused spicules are difficult to find or recognize, but skeletons composed of welded or cemented elements are common to a number of Mesozoic and Cenozoic horizons. The mineral composition of a sponge skeleton is an indication of both its classification and its lifestyle. Several forms are shown here that have characteristic shapes, but the detailed identification of many sponges relies on the study of thin sections.

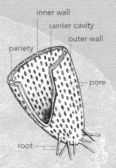

Typical features
of a sponge

Key to major groups of fossil sponges

1a Skeleton calcareous   ▶ 2
  b Skeleton composed of   ▶ 3
    siliceous spicules

2a Skeleton with thin walls;   *example:* **Peronidella**
    cylindrical in shape;   (page 60)
    small osculum
  b Skeleton with thick walls;   *example:* **Raphidonema**
    solitary or colonial; cup or   (page 61)
    vase shaped with distinct
    central cavity;

3a Skeleton cup or vase   *example:* **Ventriculites**
    shaped, rarely branched;   (page 60)
    spicule arrangement box-like,
    spicules six-rayed
  b Skeleton massive, densely   *example:* **Siphonia**
    constructed with small or   (page 60)
    reduced openings; spicules
    large, knobbly; sponges
    sometimes leave burrow or
    boring as a trace

Doryderma

Chenendopora *Cretaceous: E*
Medium-sized to large, usually 5–10 cm high. A vase-shaped sponge with a large, wide cloaca. Pores on outer and inner faces more clearly visible inside cloaca. Attachment stem shown at base.

Siphonia *Middle Cretaceous–Tertiary: E*
This genus is a member of the **demosponges**. The main body is globular, widening downward. Cloaca narrow, less than 1 cm. Surface generally smooth with small pores. Stalk long and slender; the whole skeleton having a tulip-like appearance.

Ventriculites *Cretaceous: E*
The skeleton is thin walled, vase-shaped, high to flattened and saucer-shaped (both shown below). With strong vertical grooves on the outer surface marking the course of canals and large pores on the upper face. Cloaca varying in width with shape of whole animal.

Peronidella *Triassic–Cretaceous: E*
A medium-sized form consisting of numerous cylindrical units each less than 1/2 inch in diameter and radiating from a common base. Each has a small cloaca at its tip.

2in

Chenendopora

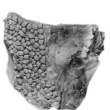

Entobia

Peronidella

Siphonia

Ventriculites

Ventriculites

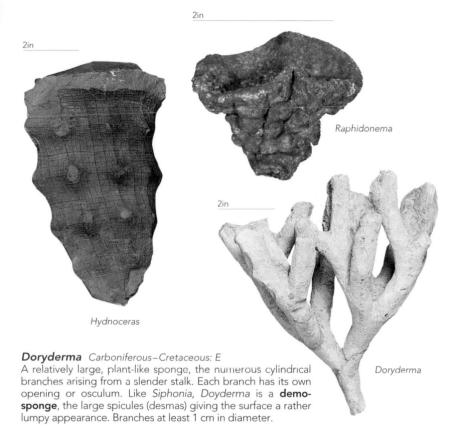

2in

2in

Raphidonema

2in

Hydnoceras

Doryderma

Doryderma *Carboniferous–Cretaceous: E*
A relatively large, plant-like sponge, the numerous cylindrical branches arising from a slender stalk. Each branch has its own opening or osculum. Like *Siphonia*, *Doyderma* is a **demosponge**, the large spicules (desmas) giving the surface a rather lumpy appearance. Branches at least 1 cm in diameter.

Hydnoceras *Devonian–Carboniferous: NA E*
Small to large (shown above). Vase-shaped. Surface with a network pattern formed by large vertical and transverse ridges with finer ridges between them. Regularly arranged swellings are present, usually at the intersection of large ridges; these swellings delimit the eight faces of the sponge. This genus represents a group that is particularly common in the Devonian of New York state.

Raphidonema *Triassic–Upper Cretaceous: E*
Rigid skeleton comprised of three-rayed spicules. Characterized by erect, vase-shaped or inverted cone-like skeletons. The outer surface is rough to the touch. Large pores are visible.

Entobia *["Cliona"] Jurassic–Recent: Worldwide*
A small, burrowing **demosponge**. The meandering galleries formed by the sponge appear as nodular swellings in shells or on rock surfaces. These swellings are joined by slender connecting rods. These are trace fossils usually preserved as casts.

CORALS

(Phylum *Cnidaria*)

Corals are marine animals with a soft, polypoid body similar to that of a sea anemone. They deposit calcareous tube-like **corallites** that combine in colonial genera to form a **corallum**. The **stony corals** are the most common fossil representatives of the *Cnidaria*. **Calcite** or **aragonite** skeletons are commonly preserved as fossils. Detailed classification is based largely on evidence obtained when specimens have been cut or thinly sectioned and examined by light and scanning microscopes. However, natural weathering and breakage of corals reveal useful details, and identifications can also be narrowed down by taking into account the age of the deposits in which the corals are found.

Key to major groups of fossil corals

1a Septa poorly developed ▶ 2
  or absent

  b Septa well developed ▶ 3

2 Horizontal partitions (tabulae) ▶ **tabulate** (page 64)
  the dominant structural element

3a Solitary or colonial corals with ▶ **rugose** (page 66)
  well-developed tabulae and
  septa confined to Paleozoic

  b Corals with developed septa ▶ **scleractinian** (page 69)
  and tabulae; confined to
  Mesozoic and Cenozoic

Isastrea
(calicinal view)

Corals in transverse and calicinal views consist either of a single star-like radial structure (corallite), as in **solitary** corals (like *Montlivaltia*, page 72) or of repeated patterns of radial structures, as in **colonial** corals (like *Hexagonaria*, page 67). The outer, youngest, or **oral** end of a corallite is the **calice** (a), occupied in life by an anemone-like **polyp**. Corallites are usually surrounded or supported by a wall or **theca** (b). If walls are missing in colonial corals, corallites consist only of "centers." Corallites may be joined by intercorallite tissue or **coenosteum** (like *Favia*, page 73). Radial elements or **septa** (c) consist of plates or spines sometimes extend beyond corallite walls as **costae** (like *Montlivaltia*). **Axial structure** (d) and/or rod- or plate-like **columella** (e) may be present or absent. Longitudinal sections reveal internal supporting structures like small blistery plates or **dissepiments** (f), and broader plates or **tabulae** (g). Colonial forms are commonly massive (head coral), or branching, encrusting, platy, tabular, or columnar. Branching forms may consist of a single corallite for each branch (**phaceloid** branching) as in *Siphonodendron* (page 68) or branches composed entirely of numerous corallites (**ramose** branching) as in *Acropora* (page 69).

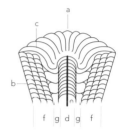

Generalized morphology of a single corallite

Favosites (calicinal view)

2in

TABULATA

Halysites: longitudinal section

Halysites: transverse section

Halysites

Ordovician–Permian
Always colonial with relatively few skeletal elements. Corallite diameters usually small, typically less than 1 to 5 mm. Walls often with pores. **Septa** absent, or very few, rarely more than 12; often as longitudinal series of spines or spinose combs. Tabulae usually well developed.

Favosites *Upper Ordovician–Middle Devonian: Worldwide*
A colonial coral with slender corallites. These are five-sided, having short, **spinose** septa and interconnecting pores. The tabulae are closely spaced and strongly developed.

Coenites *Middle Silurian–Middle Devonian: NA E Asia*
Colonies, small, delicate, **ramose branching**. Corallites very small, prismatic, in close contact, opening on surface at acute angles, with L-shaped **calices**. Walls with sparse pores. Septa or septal combs in one to three longitudinal rows. Tabulae thin, transverse to inclined.

Halysites *Late Ordovician–Upper Silurian: Worldwide*
Commonly known as the "chain coral." Colonies in transverse view have corallites arranged in chain-like ranks or **uniserial rows** (a), connected to each other in networks. Ranks palisade-like, with spaces or **lacunae** (b) between them. Corallites rounded to elliptical, with **quadrangular tubules** (c) alternating between them. Walls thick. Septa weak to absent. **Tabulae** (d) horizontal, in both corallites and tubules.

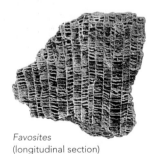

Favosites
(longitudinal section)

Coenites

2in

Halysites
(calicinal view)

Thamnopora (Pachypora) *Silurian–Permian: Worldwide*
This unusual tabulate coral forms massive, erect colonies.
These are composed of short, closely packed corallites that
branch frequently. Short, spinose septa and thin tabulae
are present.

Aulopora *Ordovician–Permian: Worldwide*
A colonial coral often found encrusting other organisms or
inorganic surfaces. The corallites are short and slightly
curved. They form a chain, with corallites cemented in "head-
to-tail" fashion. The trumpet-shaped corallites contain poorly
developed septa.

Syringopora
Upper Ordovician–Lower Carboniferous: Worldwide
?Upper Carboniferous: Worldwide ?Lower Permian: NA
The long, cylindrical, and irregularly branched corallites form
large colonies. Walls moderately thick. Septa spinose to
absent. Tabulae numerous, sagging, funnel-like.

Thamnopora
(Pachypora)

Aulopora

2in

Syringopora
(longitudinal view)

65

2in

Lonsdaelia

Lithostrotion

RUGOSA

Middle Ordovician–Upper Permian
Solitary or colonial. Corallites may attain diameters greater than 2 cm, even up to 10 cm or more in solitaries. Corallite symmetry fundamentally bilateral, though often radial in appearance. Septa usually numerous, typically in two alternating radial lengths (major and minor septa). Internal structures often include well-developed zonal arrangement of dissepiments and tabulae, with dissepiments usually confined to marginal region, or absent. Axial structures sometimes well developed, with or without a rod or plate.

Hexagonaria *Middle–Upper Devonian: NA E Asia Aust*
Colonies massive to tabular, with closely packed corallites about 1 cm in diameter. Septa long, notably thicker midway along their length, bearing small ledge-like ridges or **carinae** that are oblique to the corallites. No axial structure. Dissepiments numerous in marginal zone. Tabulae flat to subhorizontal. Calices commonly with central depression, surrounded by outer platform corresponding to dissepiment zone, edge of which is often marked by an apparent "inner wall" in transverse sections. [Genus easily and commonly confused with *Prismatophyllum*, Lower–Middle Devonian: NA]

Lithostrotion *Carboniferous: Worldwide*
This is one of the best-known Carboniferous corals. Numerous species occur worldwide, with the large to massive colonies abundant in Lower Carboniferous limestones in shallow water. The forms of the colonies vary according to the shape and arrangements of the corallites. In *L.junceum* they are circular in outline, cylindrical, and not in contact with each other. Tightly packed, four or five-sided corallites are found in the species *L.basaltiformis*. Within the outer wall, the various species are characterized by a small, rod-like, central boss. The septa are thickened and short, and the tabulae are conical in shape.

Lonsdaelia *Carboniferous: ?NA E Asia*
Although superficially similar to *Lithostrotion*, sufficient features exist to distinguish between the two in the field. Both form massive, tightly packed colonies and have angular-walled corallites. *Lonsdaelia*, however, lacks a central boss. It is replaced by a large axial structure which is surrounded by a deep circular pit. The outer walls are strong and the septa and the dissepiments well developed.

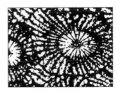

Hexagonaria: transverse section

Siphonophyllia *Lower Carboniferous: E Af Asia*
Solitary, large diameter (greater than 4 cm), sometimes very long, with numerous bends. Septa numerous, withdrawn from axial region, which appears void-like. Dissepiments numerous, in well-developed outer zone, around wide inner zone of numerous complete tabulae which are flat to slightly domed upwards in axial region and turned down in marginal region. Septa thickened here in this down-turned zone. In mature specimens, marginal zone consists only of blistery dissepiments and septa are withdrawn from wall (lonsdaleoid condition).

Siphonophyllia: transverse section

Palaeosmilia *?Upper Devonian, Carboniferous: E Af Asia Aust*
Solitary; large (greater than 5 cm). Septa very numerous; majors extending close to axis. Dissepiments numerous, in well-developed outer zone, around wide inner zone of numerous incomplete tabulae domed upwards in axial region. [Colonial forms with similar structures should be referred to as *Palastraea*, Carboniferous: NA, E Af Asia]

Palaeosmilia: transverse section

2in

Hexagonaria
(calicinal view)

Paleosmilia (longitudinal view)

Siphonophyllia
(longitudinal view)

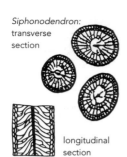

Siphonodendron:
transverse
section

longitudinal
section

Siphonodendron

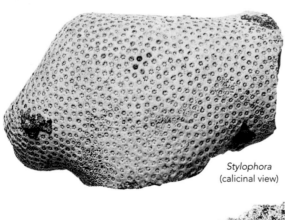

Actinocyathus: transverse
section

Siphonodendron *Carboniferous: NA E Af Asia Aust*
Colonies phaceloid, with subparallel branching cylindrical
corallites, 2–12 mm diameter. Septa strongly alternating. Columellar plate lenticular in section, sometimes absent, sometimes continuous or nearly so with septa aligned in same
plane. Dissepiments rarely absent, usually in one or a few
marginal rows, within which are well-developed flat tabulae,
pointing upward where they meet columella. [*Lithostrotion*,
Carboniferous: Worldwide, is a closely related form with very
similar internal corallite detail, but with close-packed (not
phaceloid) corallites.]

Actinocyathus *Lower Carboniferous: E Asia*
Colonies massive, tabular, with closely-packed corallites
about 1 cm in diameter. Septa withdrawn slightly from axial
region and very strongly from walls, where marginal zone
consists only of large blistery dissepiments (*lonsdaleoid*
condition). Axial structure a complex of septal elements and
small axial tabulae giving characteristic cobweb pattern.
Tabulae between axial structure and outer zone are flat to
sagging. [*Lonsdaelia*, Carboniferous: ?NA E Asia, is a closely
related form with very similar internal corallite detail, but with
phaceloid (not close-packed) branching corallites.]

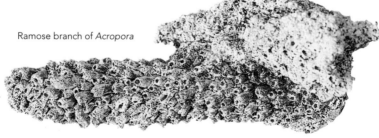

Stylophora
(calicinal view)

Ramose branch of Acropora

2in

68

Siphonodendron (longitudinal view)

2in

Actinocyathus (longitudinal view)

SCLERACTINIA

Middle Triassic–Recent
Solitary or colonial. Colonies often elaborate in form. Corallites may attain diameters of 2 cm, even up to 20 cm or more in solitaries. Corallite symmetry fundamentally bilateral, though often radial in appearance. Septa usually numerous, arranged in successively smaller size orders, usually in multiples of 6 (e.g. 6+6+12+24+48 etc). *Acropora*, *Porites*, *Favia* and *Diploria* (all shown here) are important reef builders, especially since the Miocene.

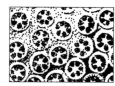

Stylophora:
calicinal view

Stylophora *Paleocene–Recent: Worldwide in lower latitudes*
Colonies encrusting, nodular and branching ramose, up to 50 cm or more across. Branches narrowly cylindrical to stoutly lobate. Corallites small, about 1 mm diameter, joined by narrow, solid **spinulose coenosteum**. Walls often project strongly (except in worn fossils), often assymmetrically as "hoods," giving a rasp-like appearance to surfaces of unworn colonies. Septa six, usually extending to axis where a slim rod may develop.

Acropora *Paleocene–Recent: Worldwide in lower latitudes*
Mostly ramose branching (e.g. bushy, stagshorn, and elkhorn forms), columnar or encrusting; table forms also common, developed by dense horizontal anastomosing of radiating branches, often with close short upturned equal branch tips. (Overall colony form of fossil specimens is rarely seen however, as *Acropora* mostly occurs as broken fragments.) Branches usually cylindrical, tapering, often long, with strong **axial corallite** at each tip, from which surrounding **radial corallites** have been generated. Corallites usually project strongly as tubules or as pustules about 2 mm across, often turned obliquely toward branch tips. Coenosteum porous, with regular pattern of very fine flaky or spinulose elements. Septa few (12, 6, or virtually absent), spine-like, deep.

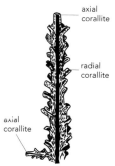

axial
corallite

radial
corallite

axial
corallite

Acropora: longitudinal
section

Septal scheme in *Porites*

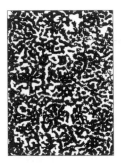

Porites: transverse view

Thamnasteria: transverse
view

Porites *Eocene–Recent: Worldwide in lower latitudes*
Colonies massive, encrusting, platy, columnar, nodular, or
ramose branching, up to several feet across. It is found
in modern reef environments. Corallites small (0.04 to 0.08in
in diameter), close-packed or joined by coenosteum, relatively inconspicuous, giving surfaces a finely pitted, cellular or
grainy appearance. Skeletal elements discontinuous, porous,
three-dimensional reticular structure of very small granulated
pillars and rods. Few septa are present in each corallite, those
that are present are short and spiny, and do not join up with
the radial structure. (Skeletal details, including corallites,
often difficult to differentiate in fossil forms, sections through
which may appear as discontinuous, subparallel, undulating
elements, or vaguely-defined centers; hence widely mistaken
for sponges.)

Thamnasteria *Triassic–Cretaceous: Worldwide*
Massive colonial coral often found in association with *Isatrea*.
The corallum is usually branched and "stick-like." Corallites
small (1–2 mm diameter), densely packed. Septa of adjacent
corallites continuous with one another (confluent); longer
septa converge at corallite centres; shorter septa curve to
join longer septa. Septa bear small ledge-like outgrowths
(**pennulae**). Longitudinal bars connect adjacent septa especially at margins of corallites where they make a wall-like
structure. Columella is a conspicuous small rod. (Confluent
septal patterns occur in numerous other corals, some of
which have been confused with *Thamnasteria*.)

Cunnolites *Late Cretaceous: NA E Asia Af*
Solitary, round to elliptical, discoidal to domed, usually 2–10
cm in longer diameter, with deep central groove and tightly-
packed septa in very numerous, poorly differentiated size
orders. Septa perforate, bearing numerous flanged structures
(**pennulae**) and connected by rods. Undersurface bears fine
surface "skin" (**epitheca**) with closely concentric ridges. (This
coral has also been widely known as *Cyclolites*.) Characteristic

Porites (showing calices
and broken section)

2in

shape is similar to the unrelated modern mushroom corals like *Fungia*, Miocene–Recent: Indo–Pacific.

Isastrea *Jurassic: Worldwide*
Colonies massive, encrusting or platy. Corallites closely packed, usually circular and domed in appearance, 3–15 mm in diameter. Septa in weakly differentiated size orders, often granular on upper edges, and with small flange-like ridges (**carinae**) oblique to corallites. Walls weak, discontinuous or absent. In some species, the septa cross the bounding walls to fuse with those of adjacent corallites. Dissepiments are numerous. Columellar are spongy and weak.

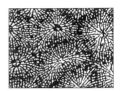

Isastrea: transverse view

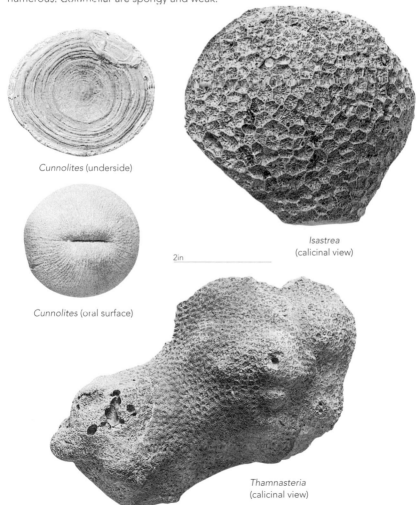

Cunnolites (underside)

Cunnolites (oral surface)

2in

Isastrea
(calicinal view)

Thamnasteria
(calicinal view)

71

Montlivaltia *Jurassic–Cretaceous–?Eocene: Worldwide*
A relatively large, solitary coral with horn-shaped to cylindrically shaped corallites. Calices with elongate axial pit, without axial structure. Septa long, in numerous size orders; may appear almost smooth in sections of larger septa, but otherwise have regular, oppositely paired, cuspate outgrowths along their length. The septa that reach to the axial area are often slightly thickened at their ends. The outer wall is rather thin and is often eroded to display the vertical septa ridges. Septa usually covered by strips of a fine surface "skin" of closely parallel ridges and grooves (**epitheca**). Dissepiments numerous, strongly arched around margins.

Placosmilia *Cretaceous–Eocene: E*
Like *Montlivaltia* but with a flattened, elongated to meandering cross-section.

Thecosmilia *Jurassic–Cretaceous: Worldwide*
Like *Montlivaltia* but in phaceloid branching colonies. Corallites robust, large (< 2cm diameter), dividing internally to form new branches.

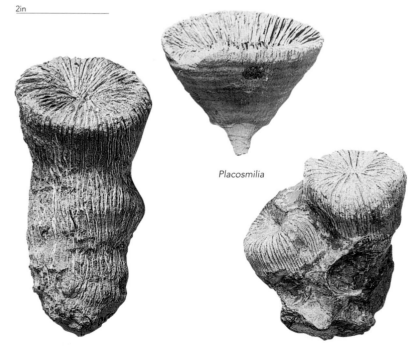

2in

Placosmilia

Montlivaltia

Thecosmilia

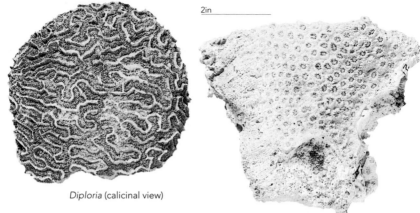

Diploria (calicinal view)

Favia *Cretaceous–Recent: Worldwide*

Colonies massive to encrusting or sometimes columnar, up to 1m or more across. Corallites 0.2 to 1 inch diameter, typically about 10mm. They are closely packed to form a massive, columnar or disc-shaped corallum. Dissepiments are well developed in this species. Septa with serrated edges and granular-spinose surfaces; projecting upwards over walls where sometimes thickened, then continuing as **costae**. Septal margins within calices steeply descending, sometimes making a crown-like structure around deep axial region. Axial structure weak to strong and spongy. New corallites arise by internal division of mature corallites.

Diploria *Upper Cretaceous –Recent: NA E Caribbean*

Characters of *Favia*, but colonies meandroid ("brain coral"). Walls very elongate, and meandering, enclosing fork-ended valley systems of numerous, barely distinguishable corallite centers along their length. Coenosteum is narrow to wide, sometimes as wide as valleys. Axial structure continuous along valley axes.

Echinopora *Miocene–Recent: Indo –Pacific*

Colonies nodular, encrusting, platy, irregularly columnar or in large delicate curving leaf-like fronds. Corallites round, projecting, 2–5 mm in diameter, with strong walls, joined by strongly spinose, tabular coenosteum. Septa in 3–4 well-differentiated size orders, and strongly spinose over well-developed wall. Axial structure large and spongy. Dissepiments numerous. New corallites arise from coenosteum between mature corallites. (This coral is similar to its well-known relative *Montastraea*, Cretaceous–Recent: Worldwide.)

Parasmilia *Cretaceous–Recent: Worldwide*

Solitary, < 10 mm diameter; narrow inverted conical form, often curved with preserved attachment structure. Each corallite is circular in section, with numerous septa characterized by a granular surface texture. Calice deep. Axial structure deep and spongy.

Echinopora (calicinal view)

Favia (calicinal view)

Parasmilia

73

JELLYFISH

(Phylum *Cnidaria*)

Other representatives of the Phylum *Cnidaria* apart from the corals occur within the fossil record. These animals include the jellyfish, hydrozoans, and stromatoporoids.

Fossil jellyfish are soft-bodied, **medusoid** creatures that are unknown other than as impressions, molds, or casts. They are, however, among the oldest known fossils, with animals remarkably similar to those that exist today recorded from the **Precambrian** Ediacara fauna (Vendian biota) of South Australia. *Mawsonites spriggi* (shown below) is a representative genus of this Precambrian fauna. The fossilization is quite unique, with the soft parts preserved as impressions on a rather coarse, sandy substrate.

Mawsonites spriggi

BRYOZOANS

(Phylum *Polyzoa*)
Colonial animals, mostly marine and important as fossils in many limestone deposits of Ordovician and later age. These typically delicate fossils may also be collected from weathered surfaces or washed from clays. Treatment of limestones with a weak solution (3 percent) of hydrochloric acid allows good quality specimens to be recovered where silicification has occurred. Each individual **zooid** of the colony builds a calcareous tube or box known as a **zooecium**, and the skeleton of the colony as a whole is called the **zoarium**. The skeleton may consist of several sizes of tubes: the larger are termed **autozooecia** (autopores), the smaller **mesozooecia** (mesopores) and **ancanthostyles**. The opening of each zooecium is known as an **aperture**. Thin sections may reveal the presence of transverse partitions (diaphragms) in the zooecia. Certain massive and thick-branched forms are very coral-like, and some groups were once treated as corals.

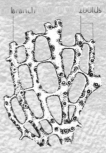

Typical features
of a Bryozoan

Key to major groups of fossil calcareous bryozoans

1a Autozooecia dominant feature   ▶ **2**

  b Autozooecia separated by intervening mesozooecia and/or ancanthostyles   ▶ **3**

2a Autozooecia rounded without frontal wall   *example:* **Berenicea** (page 81)

  b Autozooecia box-like with complex frontal wall over aperture   ▶ **cheilostome bryozoans** (not illustrated)

3a Colonies delicate   ▶ **4**

  b Colonies massive; stony   ▶ **trepostome bryozoans** (not illustrated)

4a Zooecia connected by mural pores; tubes elongate, tubular   ▶ **cryptostome bryozoans** (not illustrated)

  b No mural pores; zooecia short, cylindrical   *example:* **Fenestella** (page 76)

Fistulipora

CRYPTOSTOMATA

Apertures rectangular or polygonal, regularly arranged on colony surface. Zooecial tubes short to moderately long. Colonies erect, either flattened bifoliate fronds or narrow cylindrical branches, occasionally jointed.

Ptilodictya *Ordovician–Devonian: NA E Asia*
Fronds sickle-shaped, tapering basally. Cross-section of fronds oval or diamond-shaped with a median wall. Apertures rectangular and arranged in lines.

FENESTRATA

Apertures opening only on one side of the narrow colony branches. Zooecial tubes short. Colonies erect, often net- or fern-like.

Fenestella *Ordovician–Permian: Europe*
Net-like zoarium forming a planar fan or a funnel. The skeleton consists of a large number of radiating, slender

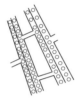

Fenestella: two rows of apertures

2in

Ptilodictya

Fenestella

2in

branches linked by thinner cross-bars or dissepiments. Each branch is pocked by the openings of numerous zooecia.

Archimedes *Carboniferous–Permian: NA Asia*
Easily identified from the spiral, screw-like axis of the colony (shown here), usually the only part to be preserved. In complete colonies the axis carries a twisted net-like frond which is virtually indistinguishable from *Fenestella* when dissociated.

Polypora *Ordovician–Permian: Worldwide*
Like *Fenestella* but having the apertures arranged in more than two rows without a central keel.

Penniretepora *Devonian–Permian: Worldwide*
Delicate, fern-like colony with primary branches bearing regularly spaced, short side branches. Apertures arranged in two rows separated by a central keel.

Polypora: more than two rows of pores

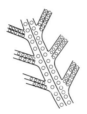

Penniretepora: surface features

2in

Penniretepora

2in

Archimedes

2in

Polypora

TREPOSTOMATA

Apertures polygonal, sometimes with two size classes, not regularly arranged on the colony surface. Zooecia tubular, long, typically thin-walled initially in colony interior but becoming thick-walled toward exterior. Colony massive, branching, frond-like or encrusting. Surface often covered with small swellings known as **monticules**.

Monticulipora *Ordovician: NA*
Growth form massive (shown here); less commonly branching or frond-like. Monticules well-developed. Zooecia having large apertures surrounded by others with small apertures. This and other "monticuliporoids" were once thought to be corals.

CYSTOPORATA

Apertures circular, usually with a hooded structure (**lunarium**) projecting over one side. Zooecia tubular, separated by areas of cyst-like calcification (**cystopores**) in most species. Colony encrusting, frond-like, massive or branching.

Fistulipora *Silurian–Permian: Worldwide*
Zoarium usually encrusting but may be massive or branching, sometimes forming large sheets up to 30 cm across (a piece of such a sheet is shown). Large zooecial apertures subcircular in shape and separated by cystopores.

Constellaria *Ordovician: NA E*
Growth form of cylindrical or flattened branches. Colony surface covered by distinctive star-shaped mounds.

Fistulipora

Monticulipora

2in

CYCLOSTOMATA

Apertures circular or polygonal, rarely semicircular. Zooecia tubular, invariably with porous walls. Large polymorphic zooecia for larval brooding present in most species.

Meliceritites *Cretaceous: E*
Zoarium consisting of slender bifurcating branches. Zooecia at colony surface are hexagonal with a semicircular aperture closed by a hinged cap (**operculum**) which may be lost.

Meandropora *Pliocene: E*
Massive zoarium consisting of radiating cylindrical bunches (**fascicles**) of tubular zooecia that branch frequently, united at intervals or linked by shelves (shown here). Polygonal apertures visible at the ends of the fascicles.

Blumenbachium *Pliocene: E*
Similar in overall colony shape to *Meandropora*, but with more complex colonies consisting of multiple layers of coalesced subcolonies which are polygonal on the colony surface and often form raised ridges at their junctions. Apertures are small and polygonal.

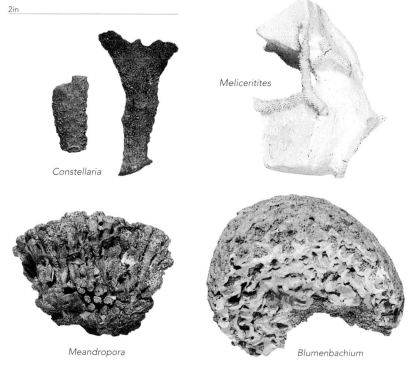

2in

Constellaria

Meliceritites

Meandropora

Blumenbachium

Stomatopora *Triassic–Recent: Worldwide*
Encrusting, thread-like, zoaria consisting of narrow, bifurcating branches, one zooecium in width. Apertures are circular and spaced regularly along the branches.

"Berenicea" *Triassic–Recent: Worldwide*
Usually found as an encrusting, colonial organism. Its colonies are mostly circular in form, the individual zooecia having thick walls and rounded apertures. They are also closely packed, and appear to radiate from a central nucleus. The specimen shown here encrusts a plate of a sea urchin.

Reticrisina *Cretaceous: E*
Erect zoarium consisting of a network of compressed branches. Apertures circular and arranged in raised rows on the sides of the branches.

CHEILOSTOMATA

This group includes the commonest living bryozoans. Growth forms are variable, and include delicate branching, net-like frondose and sheet-like encrusting zoaria. Zooecia are typically box-shaped. Apertures are usually not circular, and in living forms each aperture is closed by an operculum which is generally uncalcified. Specialized zooecia (**avicularia**) have modified opercula enlarged into mandibles used defensively.

Callopora *Cretaceous–Recent: Worldwide*
Encrusting zoaria of irregular shape with zooecia arranged in regular rows. Aperture oval and occupying most of the surface of the zooecium, ringed by small circular holes representing the bases of articulated spines. Avicularia present, visible as small diamond-shaped openings with crossbars in well-preserved examples. Specimen shown here is attached to a sea-urchin plate.

Onychocella *Cretaceous–Recent: Worldwide*
Encrusting or erect, forming sheets of variable size. Apertures approximately semicircular, located at the end of a depressed frontal wall. Zooecia commonly hexagonal in shape, often appearing slightly to overlap one another. Specimen shown here is attached to a bivalve mollusk shell.

Lunulites *Cretaceous–Recent: Worldwide*
Small zoarium, usually less than 1 cm diameter, in the shape of a low, flattened cone. Apertures of a similar shape to *Onychocella* and arranged in regular radial rows separated by grooves containing avicularia. Concave underside lacks apertures.

2in

"Berenicea"

Reticrisina

Stomatopora

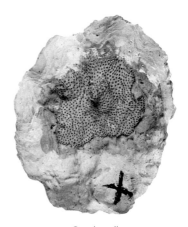

Onychocella

Callopora

Lunulites

BRACHIOPODS

(Phylum *Brachiopoda*)

Brachiopods are sea-floor dwellers. Their soft parts are protected inside a two-valved **chitinophosphatic** or **calcareous** shell. Chitinophosphatic valves are usually brown-black or black, lustrous, and with a layered horny appearance. Superficially, the shell resembles that of a **bivalve**, but the individual valves are **symmetrical** about a medial plane. They are also of **unequal size**. In contrast, the shell of a bivalve consists of two equal-sized valves that are not symmetrical.

Key to major groups of fossil inarticulate brachiopods

1a	Shell **chitinophosphatic**	▶ 2
b	Shell **calcareous**	
	(i) small with faint hinge line	▶ 2
	(ii) medium to large with distinct hinge line	▶ **articulates** (page 85)
2a	Valves elongate, subequal	▶ 3
b	Valves rounded/subrounded	▶ 3
c	Valves circular/subcircular	▶ 3
3a	Pedicle between valves	*example*: **Lingula** (page 84)

Key to major groups of fossil articulate brachiopods

1a	Hinge line straight, equal to width of shell	▶ 2
b	Hinge line curved or straight, narrower than width of shell	▶ 6
2a	Shell biconvex	▶ 3
b	Shell plano-convex	▶ 3
c	Shell concavo-convex	▶ 3
3a	Finely ribbed	▶ 4
b	Strongly ribbed	▶ 5
4	Pedicle aperture open	see **orthids** (page 86)
5a	Pedicle aperture closed	see **strophomenid** (page 88)
b	Pedicle aperture open	see **spiriferids** (page 85)
6a	Shell biconvex, ribbed	▶ 7
b	Shell biconvex, smooth or weakly ribbed	▶ 8
7a	Shell corrugated, folded with strong beak	see **rhynchonellids** (page 92)
b	Shell without fold	▶ 8b
8a	Shell subcircular to elongate tear-drop shape; prominent rounded pedicle opening	see **terebratulids** (page 91)
b	Shell strongly biconvex, beak well developed	see **pentamerids** (page 90)

Sieberella

Important features are: the **hinge line** (a); **interarea** (b); the flattened regions often present between hinge line and **beak** (c); front end or **anterior commisure** (d); the fold which is a long swelling (visible here on the dorsal valve of *Spirifer*); and the **sulcus** which is a long channel (visible page 85 on the ventral valve of *Spirifer*). The fold and sulcus often occur together on opposite valves. The ornamentation usually consists of radiating ridges or **ribs** (as in *Spirifer*); and the sulcus which is a long channel (visible on the ventral valve of *Spirifer*). The fold and sulcus often occur together on opposite valves. The ornamentation usually consists of radiating ribs (as in *Spirifer*), but concentric growth lines may also be present (as in *Atrypa*, page 85). The two valves are known as the **ventral** and **dorsal** valve. The ventral valve (e) always has the stronger beak, and is often larger than the dorsal valve (f). Also, the beak of the ventral valve often carries a small hole, the **foramen** (g), through which the attachment stalk (**pedicle**) emerges in the living animal.

Traditionally, brachiopods have been separated into two groups, Inarticulata and *Articulata*, based on whether the two valves of the shell were articulated or not. However, the most recent classification groups them into three: the *Linguliformea*, *Craniformea*, and *Rhynchonelliformea*. The valves of the first two groups are inarticulated, and those of the latter are articulated. Division into "inarticulated' and "articulated" has therefore been retained here for convenience.

Typical features of a brachiopod: side view (top) and dorsal view (bottom, not same genus)

Silicified brachiopod with preserved spiral supports

2in

Spirifer (dorsal view)

"INARTICULATE" BRACHIOPODS

The majority of inarticulate brachiopods are small with rounded, indistinct hinge-lines. Teeth and sockets are absent, as are internal support structures for the feeding organ. The shell is usually chitinophosphatic, but a number of genera have calcareous shells.

Lingula *?Ordovician–Recent: Worldwide*
Elongate and nearly oval with small, pointed hinge region. Shallowly biconvex and valves usually found separated. Ornament of numerous fine growth lines. Shell very thin with slight thickening near hinge and may have appearance of mother-of-pearl. Many species have been placed in this genus. Together, their range extends from the Lower Paleozoic to Recent, but this long range for a single genus is unlikely and these species probably represent more than one genus. True *Lingula*, as typified by *L.anatina*, probably arose in the Cenozoic.

Petrocrania *Ordovician–Devonian: E NA Asia*
Small. Usually found attached to other fossils by all parts of ventral valve, which is completely cemented to attachment surface. Shell conical or flattened and may carry radiating ribs as well as concentric growth lines. Four specimens of *Petrocrania* are shown here, arrowed, on the surface of an Ordovician strophomenid.

2in

Petrocrania (dorsal view)
attached to a strophomenid

Lingula

"ARTICULATE" BRACHIOPODS

Brachiopods with a calcareous shell. The hinge line is well developed, and internal support structures are developed in many families.

2in

Spiriferids

These are unusual brachiopods in that shells are often very wide with wing-like extensions. Spiriferids are defined by their internal spiral structure (**spiralium**) and are very variable externally. Occasionally, the spiralium may be visible on a broken or weathered specimen. Early genera are rounded in shape.

Spirifer *Carboniferous: Worldwide*
Relatively wide and strongly biconvex; hinge-line long. Wide, long interarea on ventral valve only. Beak of ventral valve strong. Strong sulcus on ventral valve and fold on dorsal valve. Ornamentation of strong ribs which fork and are present on the fold and sulcus. Growth lines may also be present. **Foramen** absent.

Atrypa (ventral view)

Eospirifer *Silurian–Devonian: Worldwide*
Biconvex but ventral valve not very deep. Beak strong and ventral valve interarea almost horizontal. Hinge-line long but less than maximum width of shell. Strong fold on dorsal valve and sulcus on ventral valve. Ornament of fine radiating ribs and concentric growth lines.

Atrypa *Silurian–Devonian: Worldwide*
Medium-sized. Early and rather untypical spiriferid. Dorsal valve very convex, ventral valve flattened or shallowly convex flexing downward at its edges. Interareas absent but hinge line long or short. Beak small and turned inward. Ornamentation of ridges crossed by equally strong growth lines. Strong fold on dorsal valve and sulcus on ventral valve, particularly in old individuals.

Spirifer (ventral view)

2in

Eospirifer (dorsal view)

2in

Athyris (dorsal view)

Platystrophia (ventral view)

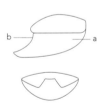

Cyrtia (dorsal view)

Cyrtia: side view (top) and front edge (bottom)

Athyris *Devonian–Triassic: Worldwide*

A small to medium-sized spiriferid with a biconvex shell. Interareas absent, hinge line short. Shape varying from wide to elongate. Fold on dorsal valve and sulcus on ventral valve, both single, smooth curves of variable strength. Beak strong and **foramen** present. Ornament of growth lines which may have the form of thick lamellae.

Cyrtia *Silurian–Devonian: Worldwide*

Medium-sized. Ventral valve (a) convex and very deep. Fold on dorsal valve and sulcus on ventral valve. Interarea of ventral valve very large (b) and almost vertical with high triangular projection in center. Shell surface smooth or carrying fine ridges and grooves.

Orthids

Hinge-line long and interareas present on both valves. Shells biconvex.

Orthis (dorsal view)

Orthis *Cambrian–Ordovician: Worldwide*

Small to medium-sized. Ventral valve convex, dorsal valve shallowly convex or flattened. Hinge-line equaling greatest width of shell. Interarea of ventral valve large; interarea of dorsal valve narrow. Interareas curve inward and both have triangular swellings or depressions near the middle. Ornament of strong radiating ribs. Dorsal valve usually with weak sulcus.

2in

Schizophoria (dorsal view)

Dalmanella (ventral view)

Platystrophia *Ordovician–Silurian: Worldwide*
Large to medium-sized. Strongly biconvex. Hinge-line may equal greatest width, produced as point or sharp corner at each end. Interareas large, almost equal in size. Beak curving inward. Strong fold on dorsal valve and sulcus on ventral valve. Ornament of radiating ribs. Externally, *Platystrophia* is indistinguishable from the spiriferids and is distinguished only by its internal structure.

Schizophoria *Upper Silurian–Permian: Worldwide*
Medium-sized. Dorsal valve more convex than ventral valve. Interarea of ventral valve larger than that of dorsal valve; interareas shorter than hinge-line which is less than greatest width. Low fold on dorsal valve and sulcus in ventral valve. Ornament of fine ribs and growth lines.

Dalmanella *Ordovician–Silurian: Worldwide*
Medium-sized. Almost circular in outline. Dorsal valve more convex than ventral valve. Interarea of ventral valve long with curved surface which slopes downwards. Interarea of dorsal valve shorter and curving upwards. Weak sulcus sometimes on dorsal valve. Ornamentation of fine ribs of variable thickness. Growth lines strong near edges of valves.

(ventral view)

(dorsal view)

Dicoelosia

Dicoelosia *Ordovician–Devonian: Worldwide*
Small to medium-sized. Strong sulci on both valves produce deep indentation on front edge. Hinge-line shorter than greatest width. The valves are sometimes smooth but frequently ornamented with fine ridges and concentric growth lines.

87

Strophomenids

Interareas present on both valves; one valve usually convex and other concave.

Strophomena: side view showing pedicle and brachial valve curvature

Strophomena *Ordovician: Worldwide*
Dorsal valve (a) convex, ventral valve (b) concave. Hinge-line long, corresponding to greatest width of shell. Interarea of ventral valve wider than that of dorsal valve. Triangular swellings in middle of upper and lower interareas. Ornament of fine radiating ribs.

Chonetes: side view showing pedicle and brachial valve curvature

Chonetes *Devonian–?Lower Carboniferous: Worldwide*
Dorsal valve (a) concave, ventral valve (b) convex. Hinge-line long but not always widest part of shell. Surface with fine radiating ribs. Interarea of dorsal valve smaller than that of ventral valve. A row of spines is present along the edge of the interarea on the ventral valve; this feature is characteristic of the group to which *Chonetes* belongs.

Rafinesquina: side view showing pedicle and brachial valve curvature

Rafinesquina *Ordovician: Worldwide*
Ventral valve shown here. Large to medium-sized. This form is like *Strophomena* but with reversed convexity, that is the dorsal valve (a) is concave and the ventral valve (b) is convex. Hinge-line long and small foramen on beak of ventral valve. Ornament of radiating ribs of variable thickness with the stronger ribs reaching to the beak. Middle rib of ventral valve usually very strong (shown here).

Strophomena (dorsal view)

Rafinesquina (ventral view)

Chonetes (dorsal view)

2in

Sowerbyella *Ordovician–Silurian: Worldwide*
Small to medium-sized. Dorsal valve concave, ventral convex. Hinge-line corresponds to greatest width of shell. Shell is semicircular in shape. Ornamentation of fine radiating ribs.

Leptaena *Ordovician–Devonian: Worldwide*
Dorsal valve concave, ventral valve convex. Hinge-line equals greatest width of shell and carries long, narrow interareas. Anteriorly the shell bends at a sharp angle to give an L shaped profile. Shells have very strong concentric ridges (rugae) and finer radiating ribs.

Productella *Upper Devonian–Lower Carboniferous: Eurasia*
Small to medium-sized, hemispherical to almost square shell with deeply concave dorsal valve (not shown here), and very convex ventral valve. Interareas very narrow, straight and poorly developed. Small spines scattered over ventral valve, but not on dorsal valve.

Spinulicosta *Devonian: Worldwide*
Small to medium-sized and similar to *Productella* to which it is very closely related. The shell is more elongate in *Spinulicosta* and carries an ornament of weak radiating ribs. Long slender spines may be present but are often not preserved. Interareas very narrow and straight as in *Productella*. Dorsal valve (not shown here) is dimpled and may carry concentric grooves.

Sowerbyella (dorsal view)

Leptaena

Productella (ventral view)

Spinulicosta (ventral view)

2in

89

2in

Sieberella
(dorsal view)

Conchidium (lateral view)

Productus
(oblique side view)

Productus
Lower Carboniferous: Eurasia N.Af China ?NA
Small to large. Ventral valve (shown here) highly convex and overlapping the hinge-line. Dorsal valve flat. Ornament of radiating ribs. Spines may be scattered over the surface and there are two rows of spines on the ventral valve near the hinge-line.

Pentamerids
Medium to large-sized. Biconvex with short, rounded hinge-line. In section, identified by the presence of organ support and muscle attachment structures. These "hang" from the dorsal valve or branch upward from the ventral valve.

Sieberella *Silurian–Devonian: NA E Af Asia*
Medium-sized and similar in general form to *Conchidium*, but with ventral valve usually even more convex. Beak very strong. **Sulcus** on dorsal valve and fold on ventral valve strong and carrying an ornamentation of ribs, but the rest of the shell surface is smooth. **Commisure** with a single, strong, angular curve.

Conchidium *Silurian–Devonian: Worldwide*
Large. Both valves very convex, ventral valve more so than dorsal valve. Beak of ventral valve curves upward and overlaps the beak of the dorsal valve (shown here in side view). Interarea of ventral valve small and interarea of dorsal valve obscured by inwardly flexed beak. Ornament of strong ribs. Fold and sulcus not developed. Commisure straight or with shallow curve.

Terebratulids

Biconvex, nonstrophic brachiopods. Rounded to elongate in shape. Interareas on ventral valves only, if visible. Shell surface usually smooth and foramen clearly visible on beak.

Dielasma *Carboniferous–Permian: Worldwide*
Small to medium-sized. Biconvex, shell surface smooth. Shell elongate, teardrop shaped. Commisure may show a single curve which may be only feebly developed as shown. Concentric growth lines occur on both valves. **Foramen** open and beak pointing upward and outward.

Dielasma: anterior view

Gibbithyris *Cretaceous: E Asia*
Medium-sized, biconvex. Commisure with double curve as shown. It may be described as inflated, and has a somewhat rounded appearance. Foramen open and beak pointing upward, or upward and inward. Shell surface smooth and marked by well-developed and closely packed growth lines.

Gibbithyris: anterior view

Ornithella *Jurassic: E Af*
Small to medium-sized, biconvex with a smooth surface and weak or strong growth lines. Outline is an elongate oval, and commisure has an upward curve which is depressed centrally. Foramen clearly visible and beak pointing upward and outward.

Sellithyris *Cretaceous: E Af*
Medium-sized. Body flattened and biconvex. Shell surface smooth with strong growth lines. Commisure complex as shown and similar to that of *Gibbithyris*. Foramen open and large. Beak pointing upward, or upward and inward.

Sellithyris: anterior view

2in

Dielasma
(dorsal view)

Ornithella
(dorsal view)

Gibbithyris (dorsal view)

Sellithyris (dorsal view)

91

Rhynchonellids
Small to medium-sized brachiopods. Biconvex and rounded, with prominent beak and coarsely ribbed ornamentation.

Goniorhynchia: anterior view

Goniorhynchia Jurassic: E
Medium-sized, biconvex, wider than long. **Commisure** as shown with single, strong, angular upward curve. **Sulcus** of dorsal valve and fold of ventral valve strongly developed. Beak strong and pointing upward and outward. Ornamentation of strong sharp-edged ribs giving the line of contact between valves a strong zigzag appearance.

Cyclothyris Cretaceous: NA E Af
Relatively large rhynchonellid, similar in general form to *Goniorhynchia* but wider and more flattened. Upward fold of commisure weaker than in *Goniorhynchia*. Beak pointing upward. It has many costae that radiate from the hinge line. The hinge-line is nonstrophic. A well-developed pedicle opening is present.

Rhynchotrema Ordovician: NA
Small, biconvex. Sulcus of dorsal valve and fold of ventral valve well developed. Zigzag commisure. Ornamentation of very strong ribs. Beak strong.

Hypothyridina Devonian: Worldwide
Large to medium-sized. Shell very high and biconvex. The commisure is characteristic, as the ventral valve projects strongly upward and meets the dorsal valve near the top surface of the shell. The sulcus on the ventral valve and fold on the dorsal valve are well developed. Ornamentation smooth near beak but strong ribs near the front.

2in

Rhynchotrema (dorsal view)

Goniorhynchia
(dorsal view)

Hypothyridina (dorsal view)

Cyclothyris (dorsal view)

Hypothyridina (anterior view)

MOLLUSKS

(Phylum *Mollusca*)
The most numerically important group of fossil animals which includes three major groups, still living today: **Gastropods** (snails), **Cephalopods** (squids, nautiluses, ammonites, etc.) and **Bivalvia** (bivalves or clams). Other groups included here are **bellerophonts**, now extinct, and **scaphopods** (tusk shells). Not included are the extinct bivalve-like **rostroconchs**, and the **Polyplacophora** (chitons or coat-of-mail shells). There are also nonshelled groups with no evident fossil record.

The mollusks occupy many ecological niches and fulfill many ways of life. The clams (bivalves) may move freely or burrow and bore into the substrate. Snails crawl and scavenge. Squids and associated forms swim and hunt in the open seas. It is likely that all mollusks evolved from a common ancestor in **Precambrian** times. The living **chitons** may be closest to this ancestor. Chitons have a segmented shell and a rather simple body structure.

Fossil mollusks are invariably recognized by their shells. These vary in structure, and tell us much about the form and habits of their original occupant. Molluskan shells are composed of calcium carbonate, either in the form of calcite or aragonite, which in many genera is covered by a thick, dark colored, organic layer called the **periostracum**. The latter rarely is found fossilized.

Key to major groups of fossil mollusks

1a Shell segmented   **chitons** (not included)
   b Shell unsegmented   ▶ 2

2a Shell coiled   ▶ 3
   b Shell uncoiled   ▶ 5

3a Shell chambered   ▶ 4
   b Shell nonchambered   ▶ **gastropods** (page 94)

4a Chamber partitions folded   see **ammonites** (page 108)
   b Chamber partitions straight   see **nautiluses** (page 106)
      or slightly curved

5a Shell composed of fused   see **scaphopods**
      single unit   (page 117)
   b Shell composed of   ▶ **bivalves** (page 118)
      two valves

Brasilia bradfordensis

BELLEROPHONTS

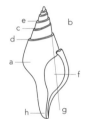

Bellerophon: aperture

Gastropod-like shells, but it is not certain that the body plan is twisted like that of a gastropod. Terms used are as for gastropods (below).

Bellerophon *Silurian–Triassic: Worldwide*
Usually 1–3 inches wide. Shell wide, flaring near **aperture** (a). Bilaterally symmetrical; last whorl covers earlier whorls, which are only visible in deep holes on either side. Front margin of aperture carries deep **slit** (s). In most specimens, the slit is gradually infilled and may be represented as a distinct ridge. Strong **ridge** (r) around middle of whorl, and growth lines strong.

GASTROPODS

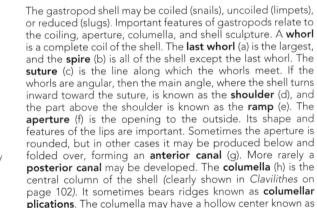

Typical structure of a gastropod as shown by *Clavilithes* (page 102)

The gastropod shell may be coiled (snails), uncoiled (limpets), or reduced (slugs). Important features of gastropods relate to the coiling, aperture, columella, and shell sculpture. A **whorl** is a complete coil of the shell. The **last whorl** (a) is the largest, and the **spire** (b) is all of the shell except the last whorl. The **suture** (c) is the line along which the whorls meet. If the whorls are angular, then the main angle, where the shell turns inward toward the suture, is known as the **shoulder** (d), and the part above the shoulder is known as the **ramp** (e). The **aperture** (f) is the opening to the outside. Its shape and features of the lips are important. Sometimes the aperture is rounded, but in other cases it may be produced below and folded over, forming an **anterior canal** (g). More rarely a **posterior canal** may be developed. The **columella** (h) is the central column of the shell (clearly shown in *Clavilithes* on page 102). It sometimes bears ridges known as **columellar plications**. The columella may have a hollow center known as the **umbilicus**. A pad of **callus** is often developed in the columellar area. The sculpture of a gastropod shell may follow the line of coiling (i.e. spiral), or it may be parallel to the growth lines (i.e. axial). In the descriptions, "width" refers to the diameter of the largest whorl, and "height" to the height of the spire.

Trepospira

Mourlonia

Poleumita *Silurian: NA E*

Coiled gastropod. Usually 2–4 inches wide. Upper surface flattened. Ornament of fine **lamellae** and slightly raised spines on shoulder. Usually angular in appearance. A similar form *Straparollus* (Silurian–Permian: Worldwide) has an aperture of different shape.

Poleumita: aperture

Trepospira *Devonian–Permian: NA SA E Af*

Usually 1–2 inches long. Conical. Deep **slit** (s) on front edge of aperture. Faces of whorls flat; outer edge of whorl sharp. Aperture as shown. Surface smooth with row of tubercles just below suture; these distinguish *Trepospira* from *Liospira* (Ordovician–Silurian: NA E Asia) which has a completely smooth surface.

Trepospira: aperture

Mourlonia *Ordovician–Permian: NA E Asia Aust*

Usually 1–3 inches long. Conical; sutures more deeply impressed than in *Trepospira*. Two to three ridges along shoulder and just above suture on earlier whorls. Strong slit on front edge of aperture (not shown here).

Mourlonia: aperture

Bellerophon

2in

Poleumita

95

Worthenia: aperture

Worthenia *Carboniferous–Triassic: Worldwide*
Medium-sized, usually about 1–2 inches high. Shell relatively higher than in *Mourlonia*. Whorls angular with flattened faces and strongly ridged shoulder bearing small tubercles. Under-surface of last whorl with spiral ridges crossed by strong growth lines, thus forming a network pattern. Aperture almost square with thickened back edge and small **slit** (s) on front margin. Umbilicus absent.

"Pleurotomaria" *Jurassic–Cretaceous: Worldwide*
Usually up to 4 inches long and/or 3 inches high. Coiling low (as shown here) to high as in *Bathrotomaria*. Umbilicus present. Aperture rounded with long slit on upper front edge (shown here just below the green spot). Heavy ornament of large swellings on shoulder of whorls and near suture. A spiral band of different sculpturing lies between the two rows of tubercles. Spiral grooves and strong growth lines also present. (*Pleurotomaria sensu stricto* is known only from the Jurassic. There are other closely similar, related forms which have different ranges.)

Bathrotomaria *Jurassic–Cretaceous: Worldwide*
Medium-sized to large, up to 3 inches high. Closely related to *Pleurotomaria* and similar to flattened or high forms of that genus. Deep slit on front margin of aperture (shown here) which becomes filled in during growth as a strong spiral ridge visible almost as far as the apex. Ornament also includes numerous spiral ridges and grooves and weaker growth lines.

2in

Bathrotomaria

"Pleurotomaria"

Worthenia

Calliostoma

Cirrus

Platyceras

2in

Platyceras *Silurian–Permian: Worldwide*
Representative of a group (Platyceratoidea) in which the last
whorl is very large and the other whorls are much smaller. The
specimen shown is extreme, and other members of the group
may be more similar in general form to *Mourlonia* (page 94).
Border of aperture may be wavy or straight. Ornament of
growth lines. No slit on front margin of aperture.

Calliostoma *Cretaceous–Recent: Worldwide*
Medium-sized, usually 0.5–2 inches long. Conical with pointed, straight-sided spire. Aperture as shown. Umbilicus
absent. Inner shell layer commonly like mother-of-pearl
(shown here). Sutures may or may not be deeply indented.
Ornament of spiral ridges, varying in distribution from area
near sutures only, to covering whole surface of whorls, also
varying in strength. No slit on margin of aperture.

Calliostoma: aperture

Cirrus *Triassic–Jurassic: SA E*
Medium-sized to large, 1–3 inches wide or high. Flattened to
high conical (shown here). Umbilicus large, varying with
height of shell. No slit on margin of aperture. Ornamentation
of strong vertical ridges and weaker spiral ridges. Sutures
shallowly depressed. Aperture almost circular. Coiling opposite
in direction to most gastropods.

97

Ooliticia *Jurassic–Cretaceous: Worldwide*
Small to medium-sized, usually 0.25–2 inches high. Steeply conical with faces of whorls rounded. Umbilicus absent. Aperture diamond-shaped to rounded. Ornament of strong spiral ridges carrying tubercles and crossed by fine vertical ridges. No slit on margin of aperture.

Loxonema *Silurian: NA E*
Medium-sized up to about 3 inches high. High, pointed spiral with whorls with rounded walls. Sutures deep. Umbilicus absent. No slit on margin of aperture but outer margin with a deep curved depression known as a **sinus** (s). Ornamentation absent.

Loxonema: aperture

Microptychia *Carboniferous: NA E*
Medium-sized, up to 3 inches long. High pointed conical. Sutures deep with ornament of short vertical ridges; these increase in strength upward and may completely cover the top whorls. Lower whorls smooth. Aperture almost circular and lacking sinus. Walls of whorls rounded but more convex near lower suture.

Natica *Paleocene–Recent: Worldwide*
Medium-sized, shell-drilling, predatory gastropod, usually 0.5–2 inches high. Shape ranges from almost spherical (shown here) to conical. Walls of whorls rounded and sutures usually deep. Surface smooth and may be shiny with a few lamellar growth lines near aperture. Umbilicus usually present but **columellar** callus may cover it. Last whorl very large. Aperture oval to circular. Inner lip thickened, outer lip thin.

Ooliticia

2in

Natica

Loxonema

Microptychia

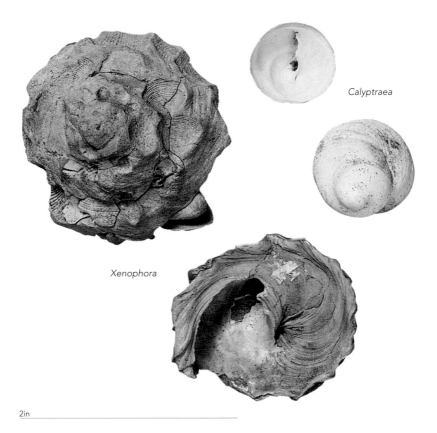

Calyptraea

Xenophora

2in

Xenophora *Cretaceous–Recent: Worldwide*
Medium-sized, up to 3 inches wide. Conical with flattened base. Last whorl with sharp outer margin. Wide umbilicus and characteristically shaped aperture (shown here). Inner margin thickened. Surface rough with depressions where shell fragments, and other foreign particles such as pebbles, were attached during life. Some shell fragments are still present on the right side of the specimen, and depressions on the upper side show the patterns of attached particles. Some species of *Xenophora* have lightly sculptured surfaces.

Xenophora: cross section showing flattened base

Calyptraea *Cretaceous–Recent: NA SA E*
Medium-sized, up to 3 inches wide. Flattened to high conical shell consisting of a few wide whorls. Last whorl very large and lower surface deeply concave with a small internal shelf which has a twisted border (columella) and a small umbilicus at its highest point. Ornamentation of weak growth lines and occasional tubercles which are stronger near the lower edge.

Calyptraea: cross section showing concave lower surface

Crepidula: cross-section

Crepidula (slipper limpet) *Tertiary–Recent: NA E*
Medium-sized, usually 1–3 inches long. Flattened, convex and slipper-like. Whole shells consist of a single whorl. Undersurface characteristic, having deep concavity as shown, and a wide concave **shelf** (s) that lacks the thickened columellar edge of *Calyptraea* (page 99). Ornamentation of ridges, spines and growth lines may be present on upper surface.

Stellaxis

Stellaxis (sundial shell) *Eocene–Recent: NA E Asia*
Small to medium-sized, up to 1 inch across. Flattened or slightly domed with a large, wide umbilicus which is strongly sculptured, often with prominent notches. Outer margin of last whorl sharp with a spiral ridge. Sculpture of a few spiral ridges above and below suture. Aperture sub-triangular and thickened at two outer angles. (*Stellaxis* can be confused with the related genus *Architectonica*, whose whorl surfaces are covered with widely-spaced spiral grooves.)

"Aporrhais" *Cretaceous–Recent: Worldwide*
Modern range is restricted to North Atlantic. Large to medium-sized, up to 5 inches high. Turretted with long anterior spine or elongate canal. Outer lip of aperture flared, often spinose, wing. The **apical angle** is narrow and the whorls rounded in outline. A bold sculpture of vertical ridges and strongly developed tubercles may be present. Fine growth lines on outer lip and wing. Shallow burrower.

Crepidula

2in

"Aporrhais"

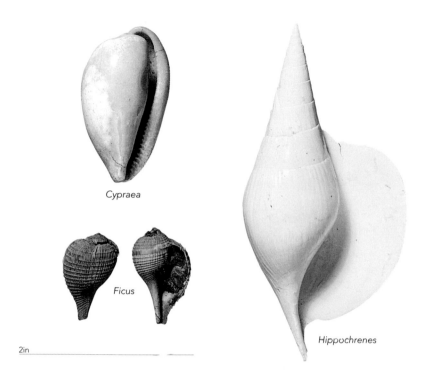

Cypraea

Ficus

Hippochrenes

2in

Cypraea (cowry) *Tertiary–Recent: Worldwide in warm waters*
Small to large, 0.25–6 inches long. Highly characteristic
conch-shape with outer lip of last whorl greatly expanded
and completely covering the rest of the shell. Aperture is
long, with thickened outer lip serrated. Surface usually
smooth and shiny.

Ficus (fig shell) *Paleocene–Recent: NA E Asia*
Small to large, usually 0.5–5 inches long. Low spired, shell spin-
dle-shaped (**fusiform**) with very large last whorl and lower region
produced as broad, twisted canal. Whorls of spine rounded or
with shoulders. Aperture large broad and elongate. Shell thin
with little columellar callus. Sculpture of spiral and axial ribs.

Hippochrenes *Eocene: E Asia*
Medium-sized to large with tall spire, approximately equal to
height of last whorl. Lower part of last whorl produced as
elongate canal. Outer lip expanded as a large flare which is
fused to the spire. The lower face of the flare carries a deep
groove along the junction with the spire, and the shell below
this groove may be expanded to cover it. The specimen shown
has moderate development of the flare, but in some forms it
may resemble that of *Aporrhais*.

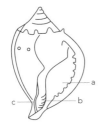

Galeodea: showing features

Volutospina: aperture region

Galeodea *Eocene: NA E Asia*
Medium-sized, usually 1–3 inches long. Like *Ficus* (page 101) but with higher, conical spire. Whorls angular with strong spiny projections at shoulder; spiral ridges and more swellings of variable strength on last whorl. **Aperture** (a) elongate, with thickened outer margin carrying serrations. Strong **columellar callus** (c) with several strong ridges on **inner margin** (b), especially at lower end.

Volutospina *Paleogene: NA E Af Asia*
Medium-sized, usually 1–4 inches long. Representative of a group in which the spire is of intermediate height. Whorls angular, usually carrying ribs with spines at the shoulder. **Aperture** (a) narrow with short **canal** (b). Columellar **plications** twisted (c).

Marginella *Eocene–Recent: Worldwide*
Small to medium-sized, less than 0.25 inches to 2 inches long. Often tapering equally at each end, oval or elongate. Surface smooth, unsculptured. **Aperture** (a) elongate (shown here), outer margin thickened and sometimes bearing teeth (not shown here). Several columellar **plications** present (b,c,d). Sutures slightly impressed.

Clavilithes *Eocene–Pliocene, ?Recent: NA E Asia*
Medium-sized to large, usually 4–6 inches long. Shell elongate, conical with sutures deeply impressed. Spire short, pointed and often strongly sculptured at the apex; lower parts of shell smooth. Broad, almost flat, ramp above shoulder; rest of whorl almost vertical. Whorls increasing uniformly in size. Long canal. Aperture as shown *(see also* page 94). No columellar plications. A longitudinal section of the shell shows the ramp, whorl shape, canal, aperture, and columella.

2in

Galeodea

Volutospina

Marginella

Buccinum

2in

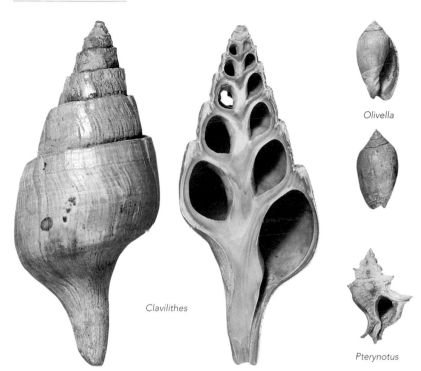

Olivella

Clavilithes

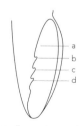

Pterynotus

Pterynotus *Eocene–Recent: Worldwide*
Usually 1–3 inches long. Dominant ornament of three strong axial ribs per whorl, each bearing a spine and with one smaller tubercle between each, overriden by spiral ridges. Aperture small. Outer lip expanded into rib as shown and having ridged inner margin. Inner lip thickened. Canal is medium to long. Columellar plications are absent.

Buccinum (whelk) *Pliocene–Recent: NA E*
Medium-sized to large, usually 1–6 inches high. Fusiform shell with whorls increasing uniformly in size. Aperture wide and oval with short canal. Ornament of spiral and/or axial ribbing. Outer lip sharp, sometimes recurved. Columellar callus relatively weak.

Olivella *Tertiary–Recent: Worldwide*
Usually less than 2 inches long. Last whorl very high in relation to rest of shell. Low spire. **Aperture** (a) elongate with short, wide **canal** (b) and **notch** at upper end (c). Columellar **plications** are present (d). Outer lip thin and sharp. Very weak axial grooves may be present. Several spiral grooves are usually developed near the lower end of the last **whorl** (e).

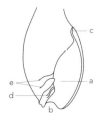

Marginella: aperture region

Olivella: aperture region

103

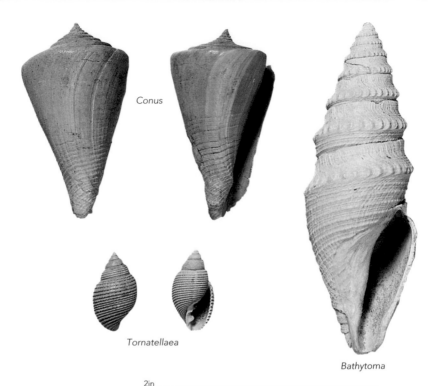

Conus

Tornatellaea

Bathytoma

2in

Tornatellaea: aperture
region

Trochactaeon: aperture
region

Conus (cone shell) *Eocene–Recent: Worldwide*
Small to large, usually 1–4 inches long. Steep, upturned, conical
shape below, with flat to steep conical spire above. Aperture
parallel-sided, long and narrow (shown here) with a notch at the
upper end. Canal short and outer lip thin. Ornament of spiral
grooves, ridges or tubercles. Spiral ridges of variable strength
on spire.

Bathytoma *Tertiary–Recent: Worldwide*
Small to medium-sized, usually 0.5–3 inches long. Shell narrow,
equally conical at both ends. Last whorl about half the total
height. Aperture elongate, almost parallel-sided. **Columellar
plications** absent. Sutures deep. Growth lines flexed backward
at shoulder with row of strong tubercles along shoulder. Spiral
ribs present.

Tornatellaea *Jurassic–Oligocene: Worldwide*
Small, usually less than 0.25–1 inches long. Spire whorls rounded. Aperture elongate, almost parallel-sided. Columellar plications absent. Sutures deep. Growth lines flexed backwards at
shoulder with row of strong tubercles along shoulder. Spiral
ribs present.

Planorbis

Trochactaeon

2in

Trochactaeon *Cretaceous: Worldwide*
Large to medium-sized, usually 1–3 inches long. Spire low, concavely pointed, body whorl large. Aperture elongate, parallel-sided. Columella with two or three strong folds at the lower end. Shell smooth and thick. In a closely related form *Actaeonella* (Cretaceous: NA), the last whorl is expanded and covers the spire, giving it an appearance superficially similar to that of *Cypraea*.

Planorbis *Oligocene–Recent: E Af Asia*
Small to medium-sized, usually less than 0.25–2 inches long. Flattened spiral (planispiral) shell with upper side flattened (shown here) and **lower side concave** (a), or with broad umbilicus and sutures deeply impressed (shown here). A low **spire** is sometimes present (b), or both surfaces may be **concave** (c). Aperture oval to wide crescentic. Outer margin sharp. Ornament of fine growth lines only.

Planorbis: cross sections showing grown forms

105

CEPHALOPODS

Squid, octopus, cuttlefish, and nautiluses are living cephalopods – mollusks which have adopted an active, predatory lifestyle, and typically reach sizes far larger than any other invertebrates (the modern giant squid *Architeuthis* may weigh up to 1000 kilograms). Cephalopods are characterized by a hollow shell, divided into chambers by **septa**. As the animal grows, these chambers are emptied, reducing the density of the animal, allowing it to become naturally buoyant. The nautiluses and ammonites are common fossils in the Palaeozoic and the Mesozoic, and are widely used for dating rocks (biostratigraphy).

Nautiluses

The most primitive cephalopods, possessing external shells either straight ("orthocones") or coiled, of which two genera *(Nautilus and Allonautilus)* are still alive today. In most nautiluses, the septa are simple disks and quite widely spaced. Most diverse in the Palaeozoic, the modern species are relatively deep-water scavengers and stay close to the sea floor.

Orthoceras Ordovician: Worldwide
Straight, cyclindrical shell that expands slightly with growth. The **siphuncle** is central and remains empty. **Cameral** deposits made of calcium carbonate are found in the **apical** part of the shell. These acted as a counterweight for the body of the animal at the front of the shell and allowed the nautilus to swim horizontally. Mostly of moderate size (less than one meter) but some as much as three meters in length.

Orthoceras

2in

Aturia

2in

Eutrephoceras

Nautilus

Endoceras

Endoceras *Ordovician: NA E Asia Australia*
Straight shelled forms with very broad siphuncles usually displaced ventrally. Some species are extremely large, more than nine meters in length.

Eutrephoceras *Jurassic–Miocene: Worldwide*
A wide-ranging, compact, medium-sized nautilus with a coiled shell ornamented with short ribs emerging from the umbilicus but otherwise smooth.

Aturia *Paleocene–Miocene: Worldwide*
Completely smooth and strongly laterally compressed, this nautiloid is characterized by closely packed, fluted septa similar to those of the *Goniatites* (see page 108).

Nautilus *Oligocene–Recent: E Aust Malay Archipelago*
The living *Nautilus* ("pearly nautilus") has a smooth shell in which the body chamber is very large. It covers or overlaps a number of earlier whorls, giving the shell an **involute** appearance. Septal sutures are gently flexed or folded appearing broadly lobed. The shell of *Nautilus pompilius* is sectioned here to show body chamber, septa and siphuncle.

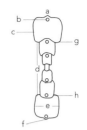

Typical structure of an ammonite: cross section (top) and longitudinal section (bottom)

Ammonites

Not known after the Cretaceous, ammonites are important fossils for dating late Palaeozoic and Mesozoic rocks.

Ammonite shells appear similar to some gastropod shells, but their shells are divided by **septa** and they have **suture** lines and a **siphuncle**. Most ammonites consist of **whorls** coiled in a plane spiral, but forms coiled in a helical spiral, as well as straight and hooked, also occur. Whorls are often ribbed and sometimes bear spines which may be broken off, leaving only a series of rounded bases (**tubercles**). The suture line is the junction between a **septum** and the outer shell, and is visible as a wavy, frilled line on the surface. Diagrams trace the line of the suture from the **venter** (a) to the **umbilical-seam** (h), where adjacent whorls join. The arrow on a suture-line diagram points toward the **aperture** (k); **lobes** are the backward folds of the suture-line and **saddles** are the forward projections. A simple suture-line is shown at the junction of the two colors on *Ceratites* (page 109), and a complex suture-line is shown on *Phylloceras* (page 109). The **siphuncle** (b) is a tube running through the septa and chambers, and is often indicated by sharp flexions of the suture-lines. The **umbilicus** (d) is the depression produced by coiling and the increasing size of the whorls, and the **umbilical shoulder** is where the shell turns inward (g) to the seam from the **lateral surface** (c). The **keel** (f) is a ridge which may be present along the venter. The thickness is (e), and arrow (j) points backward toward the aperture. Most specimens shown here are **internal molds**. Suture-lines are important for identification in some groups. Early ammonites have simpler suture lines (like *Goniatites* and *Ceratites*), and later forms have more complex sutures.

Model Ammonite

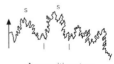

Ammonitic suture showing saddles (s) and lobes (l)

Ceratitic suture showing saddles (s) and lobes (l)

2in

Gastrioceras

Ceratites (with painted sutures)

Goniatites

Goniatites: cross section of whorl (top) and suture diagram (bottom)

Gastrioceras: cross section of whorl

Palaeozoic ammonites

Goniatites *Lower Carboniferous: Northern Hemisphere*
Small to medium. Thick, globular shells with narrow umbilicus. Angular suture-line is characteristic of the group (Goniatitida), which ranges through the Carboniferous and Permian. In this type, the ventral lobe is pointed and the lateral lobe is smooth and convex.

Gastrioceras *Upper Carboniferous: Worldwide*
This is a member of the same superfamily as *Goniatites*. Globular whorls with narrow umbilicus (shown here), but some forms are flattened with wider umbilicus. Suture-line simple with smoothly curved saddles and pointed lobes. Ornamented with strong ribs on the inner margins, which divide on the side of the whorl then extend as fine ribs across venter, flexing slightly backward in the center.

Mesozoic ammonites

Ceratites *Triassic: E*
Moderately evolute whorls and wide umbilicus. Shape of
suture-line is known as **ceratitic** and is characteristic of the
group (Ceratitida) which ranges through the Triassic. Has a
box-like cross section. Has smooth, simple rounded saddles
and serrated lobes. Forms similar to *Ceratites* are known
from the Triassic of North America.

Ceratites: cross-section of
whorl

Lytoceras *?Upper Triassic, Lower Jurassic–Upper Cretaceous: Worldwide*
Evolute shell in which the whorls enlarge rapidly toward the
aperture. The whorls have a rounded or subrounded cross section. Suture-line complex with ventral lobe divided, two lateral
lobes and edges of all lobes extremely subdivided and fern-
like. Ornamented with fine ribs. If the test is preserved, characteristic frilled ribs are present at intervals (not shown here).

Phylloceras *Upper Triassic–Upper Cretaceous: Worldwide*
Involute shell in which outer whorl covers the preceding one.
Shell compressed with small umbilicus. Suture-line complex and
shown at junction of red and white paint on specimen shown;
saddles have characteristic leaf-like projections and lobes have
pointed projections. Surface smooth or ornamented with fine
striae extending unbroken across the venter.

Phylloceras: cross section
of whorl

Lytoceras

2in

Phylloceras (with painted sutures)

109

Arnioceras

Asteroceras

Amaltheus

Dactylioceras

Promicroceras

2in

Arnioceras: cross section
of whorl

Jurassic ammonites Early Lower Jurassic

Arnioceras *NA SA E Af Asia*
Evolute whorls with very wide umbilicus. Ornamented with strong, simple ribs which bend forward near venter. Keel strong and bordered by furrows on each side.

Asteroceras *NA E Asia*
Medium-sized. Moderately evolute, thick whorls that increase in size rapidly, with wide umbilicus, and keel on venter bordered by shallow depressions. Ornamented with strong ribs that fade on the venter. Suture relatively simple (shown here).

Promicroceras *E*
Small, up to 2 inches diameter. Very evolute whorls, with circular cross section, wide umbilicus and rounded, unkeeled venter. Ornamented with strong ribs sometimes flattened where they cross the venter.

Jurassic ammonites Middle Lower Jurassic

Amaltheus *Northern Hemisphere*
Small to medium. Involute, flattened whorls, with fairly narrow umbilicus and strong keel on venter. Ornamented with strong, slightly S-shaped ribs that bend forward and fade on outer part of whorl, then strengthen again to form the characteristic serrations on the keel. Some species may have tubercles on the side walls.

Amaltheus: cross section of whorl

Jurassic ammonites Late Lower Jurassic

Dactylioceras *Worldwide*
Possibly one of the best-known of all ammonites. Evolute whorls with almost circular cross section and wide umbilicus. Ornamented with strong ribbing, in some species the ribs are simple but in others they divide over the venter. Tubercles occur on the ribs at the point of division in some forms.

Harpoceras: cross section of whorl

Harpoceras *NA SA E Af Asia*
Involute, flattened whorls, with small to medium-sized umbilicus and strong ventral keel. Ornamented with strong flexuous or sickle-shaped ribs, which bend backward in the middle of the whorl then swing forward on the outer part and extend on to the venter.

Hildoceras *E Af Asia*
Medium-sized ammonite with flattened, evolute shell. Whorls have rather square cross section and a wide umbilicus. It is characteristically grooved near middle of whorl side. Ribs absent or feeble on umbilical side of groove, but strong backwardly curved ribs on ventral side of groove. Strong ventral keel bordered by deep furrows.

Hildoceras: cross section of whorl

2in

Hildoceras

Harpoceras

111

Jurassic ammonites Middle Jurassic

Parkinsonia *E Af Asia*
Evolute whorls, with medium to wide **umbilicus**. Ornamented with sharp ribs that **bifurcate** and bend forward near the edge of the venter, then end at the edge of a deep groove in the middle of the **venter** which becomes shallower at larger sizes.

Stephanoceras *Worldwide*
Evolute whorls with almost circular cross section and wide umbilicus. Strong ribs on the inner part of the side of the whorl divide into two or three at a midlateral **tubercle**, then continue across the venter with a midventral interruption.

Graphoceras *E Af Asia*
Involute, flattened whorls with a medium to small umbilicus and a strong keel on the venter. Ornamented with flexuous ribs, similar to those of *Harpoceras*, that diminish in strength at larger sizes. The specimen shown here is complete with most of its shell, and growth lines following the shape of the ribs are visible near the aperture.

Graphoceras: cross section of whorl

Jurassic ammonites Upper Jurassic

Cardioceras *NA E Asia*
Moderately involute whorls with triangular cross section, moderate to narrow umbilicus and keel on venter. Strong ribs divide into several smaller ribs at middle of whorl side, then curve forward to form forwardly pointing chevrons (serrations) on the keel. Ribs are fine on early whorls, becoming more widely spaced at larger sizes. (The specimen shown here is the subgenus *Scarburgiceras*, with less distinct keel and less differentiated ribs.)

Cardioceras: cross section of whorl

Perisphinctes *E Af Asia*
Some specimens attain very large sizes. Very evolute whorls with almost square cross section and steep umbilical wall. The inner whorls are ornamented with many closely spaced ribs, while those of the outer whorl are separated and rather massive.

Pavlovia *E Asia*
Evolute whorls with whorl cross section broadest near edge of umbilicus. Sharp, wiry ribs divide into two near middle of side of whorl and are continuous across venter.

Pavlovia: cross section of whorl

2in

Pavlovia

Graphoceras

Parkinsonia

Stephanoceras

Cardioceras (Scarburgiceras)

Perisphinctes

113

Cretaceous ammonites Lower Cretaceous

Hamites: outline of complete shell (shading highlights the final living chamber of the mature ammonite)

Hamites *Worldwide*
Uncoiled with characteristic shape as shown, usually with straight shell between two or three hooks, but some forms have helical coiling. Many specimens are fragmentary, and only the shaded portion is shown here. The whorl has a circular or oval cross section, and is ornamented with ribs that entirely encircle the shell. Suture-line is simple.

Hoplites *E Asia*
Involute shell which is compressed laterally to give a trapezoidal cross section. Medium-sized, involute whorls, with deep, narrow **umbilicus**. **Tubercles** at edge of umbilicus give rise to prominent ribs that quickly divide and curve forward at the edge of the venter, which has a smooth central depression.

Mortoniceras *NA SA E Af Asia*
Large, flattened, evolute whorls, with a wide umbilicus, quadrangular whorl section and a strong keel bordered by furrows on venter. Ornamented with strong ribs that bear tubercles near edge of umbilicus. Most ribs divide into two on side of the whorl then curve forward up to the edge of the **venter**. Ribs are often serrated on outer part of whorl side.

Douvilleiceras *NA SA E Af Asia*
Globular, swollen whorls, evenly rounded in cross section, with medium to narrow umbilicus. Strong, single ribs bear many tubercles on side of whorl and venter, but are interrupted along the midline of venter.

Hoplites: cross section of whorl

2in

Hamites

Hoplites

Douvilleiceras

Mortoniceras

2in

Oxytropidoceras

Scaphites

Placenticeras

Turrilites

Douvilleiceras: cross section of whorl

Oxytropidoceras: cross section of whorl

Oxytropidoceras *NA SA E Af Asia*
Involute, flattened whorls with narrow, deep umbilicus and strong keel on venter. Ornamented with numerous flexuous ribs curving forward at their ventral ends.

Placenticeras *NA E Af*
Involute, flattened whorls with narrow, deep umbilicus and raised, flattened (tabulate) venter. Ribs weak or absent, but pointed tubercles border the umbilical edge; another row of tubercles on outer part of whorl, and small tubercles on side of flat venter in some forms.

Cretaceous ammonites Lower–Upper Cretaceous

Turrilites *NA E Af Asia*
Coiled in a helical spiral like some gastropods (from which they are distinguished by possessing **septa** and suture-lines). Ornamented with coarse ribs and tubercles. Suture-line is complex (not shown here).

Scaphites *Worldwide*
Rather odd shape. Normal spiral whorls are followed by a short straight section then a hook-shaped living chamber. Ornamented with many fine branching ribs and tubercles; the ribs curve forward and are continuous across the venter.

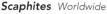

Placenticeras: cross section of whorl

Scaphites: cross section of whorl

115

Baculites: cross section
of whorl

Acanthoceras: cross
section of whorl

Cretaceous ammonites Upper Cretaceous

Baculites *Worldwide*
Up to 80 inches long. Only the very earliest (smallest) part of the shell is coiled, and is rarely seen in collections. The remainder of the shell is a single, long, straight shaft, usually found as fragments. Cross section flattened or oval, and suture-line complex. Shell smooth or ornamented with sinuous ribs.

Acanthoceras *NA E Af Asia*
Evolute, robust whorls with wide **umbilicus**, quadrate whorl cross section and keel on venter. Ornamented with straight ribs bearing tubercles at edge of **venter**.

Hoploscaphites *NA E SA Antarctica*
This genus is clearly related to *Scaphites* (page 115). The initial whorls overlap, but the last is hook-like at the end of a very short shaft. The shell is rather flat-sided with distinct ridges and **tubercles**.

Belemnites
An extinct cephalopod group, superficially like modern squid. Unlike ammonites, the shell was internal. The bullet-shaped back part of the shell, the **guard**, is dense and durable, and is a common fossil in the Jurassic and Cretaceous. A hollow region at the front of the guard, the **alveolus**, houses the cone-shaped chambered part of the shell – the **phragmocone** (shown in *Cylindroteuthis*, page 117). In complete specimens

2in

2in

Hoploscaphites

Baculites   Belemnitella

Acanthoceras

116

(rarely preserved), a flattened region, the **pro-ostracum**, is an anterior projection of the phragmocone which lies over the mantle cavity. Broken specimens of the guard show a structure of radiating calcite fibers and concentric growth lines.

Neohibolites *Upper Cretaceous: Worldwide*
Small, guard usually 2–4 inches long. Circular cross section. From its widest central part, the guard narrows toward the phragmocone, which is preserved but crushed in the specimen shown, and seen at the front end. A short slit and groove are present on the guard at the back, near the alveolus, as shown here.

Belemnitella *Cretaceous: NA E Asia*
Large, guard usually more than 4 inches long. Cross section almost circular but has a flattened upper surface with a pair of shallow, longitudinal depressions. The sides of the guard may have a granular appearance. There is a long slit on the ventral surface of the alveolus near its edge. Good specimens may onhibit traces of blood vessels over the outer surface.

Cylindroteuthis *Jurassic–Early Cretaceous: NA E*
Large, guard up to 6 inches long. Cross section oval with slightly flattened sides, and long groove on the ventral surface. The latter tends to deepen toward the tip of the guard. The chambered phragmocone is clearly shown here at the front of the specimen.

Neohibolites

2in

SCAPHOPODS (Tusk shells)

A major group of mollusks of the same status (class) as the gastropods, cephalopods and bivalves, but less common in fossil and Recent faunas, and more uniform in appearance. Shells elongate, conical, open at both ends and curved like an elephant's tusk. In life, the concave side is upward; the smaller (**apical**) opening is posterior, and the larger opening is anterior, this being buried in the sediment.
Forms with numerous ribs (e.g. *Prodentalium*, Carboniferous –Recent) acquired an apical slit during the early Tertiary (e.g. *Fissidentalium* shown here). This is the central stock of the ribbed scaphopods, and *Dentalium*, symmetrical and with few ribs, which is now dominant, did not appear until the Miocene.

Fissidentalium *Paleocene–Recent: Worldwide*
This genus has up to 40 unequal ribs asymmetrically arranged around the tube. The underside of the apex usually has a long slit, which is absent from the earlier *Prodentalium*. (The genus *Dentalium*, Miocene–Recent, has 6–16 primary ribs symmetrically placed, and an apical notch on the underside).

Cylindroteuthis

Fissidentalium

BIVALVES

Cockles, scallops, razor clams, oysters, mussels and clams are all bivalve mollusks. Bivalve mollusks superficially resemble brachiopods (see page 82), but closer inspection reveals important differences. In most bivalve mollusks, each valve is asymmetrical (**inequilateral**) with the beak toward the front end; and the valves are mirror images of each other (**equivalve**). Oysters are well-known exceptions to this: they are inequivalve. In brachiopods, each valve is usually symmetrical but the two valves differ in size and curvature.

Typical features of a bivalve shell: side view (top) and cross section

Important features of bivalve mollusks are height (h), length (l), thickness (t), the **beak** (b), and ornamentation. A flattened region between the beaks of paired valves is an **area** (shown in Arca on page 123); a flattened depression in front of the beak is a **lunule**, and behind the beak is an **escutcheon**. An opening or notch between or behind the beaks is a **ligamental notch**. Valves articulate at the "hinge line" (dorsal margin). In some shells, the valves do not meet at the front or back and a **gape** remains (shown in Pholadomya on page 120).

Inward projections of shell known as **hinge teeth**, collectively referred to as **dentition**, may be present below the beak and at either end of the dorsal margin; ridges or "teeth" on the side and lower margins are termed **crenulations** (shown in Glycymeris, page 123). Front and back muscle scars may be present (shown in Mya, page 119), or only one may be developed. The **pallial line** is a curving linear mark joining the front and back muscle scars indicating the extent of attachment of the animal within the shell. The **pallial sinus** is an inflexion of this line near the back of the shell. The shell exterior is often sculptured ("ornament") with features such as concentric ribs, radial ridges and spines, according to mode of life.

2in _____

Venericor

Arctica

Venericor *Paleocene–Eocene: NA SA E Af*
Ranging from 1–6 inches long. Equivalve and strongly convex.
Strong beak points forward. Ligamental notch behind beak.
Two strong **teeth** (a, b) under beak on each valve as shown.
Ornament of wide radiating ridges and concentric lamellae
strongest near margin. Margins with small crenulations.

Venericor: showing teeth

Arctica *Paleocene–Recent NA E*
Usually 1–4 inches long. Shape similar to *Venericor*. Ligamental notch deep. Two or three **teeth** present as shown. Pallial
sinus absent. Ornament of weak concentric ridges. Margins
lacking crenulations.

Arctica: beak showing
teeth

Orthocardium *Paleocene–Eocene: E*
Representative of group which includes cockles. Medium-
sized. Valves almost symmetrical; beak points slightly
forward. Dorsal margin almost straight. Two central teeth on
each valve, one side tooth at front and back on left valve; two
front and one back on right valve. Ornament of strong frilly
ribs with headed edges. Margins with strong crenulations.

Mya *Oligocene–Recent: NA E Asia*
Usually 1–6 inches long. Elongate, flattened. Beak small, pointing upward. Ornament of concentric lamellae or smooth. Dorsal
margin curved, lacking teeth but having a spoon-like projection
known as the **chondrophore**. Crenulations absent. Wide posterior gape. Front muscle scar high and curved; back muscle scar
circular and deep. Deep pallial sinus.

Mya

Orthocardium

Mya

2in

Pitar

Teredo

Neocrassina

2in

Teredo Eocene–Recent: Worldwide

Representative of a group of mollusks most commonly known from their **borings** in wood (shown here). Borings are circular in cross section and may have a calcareous lining. They may be filled with mud or contain shell remains. *Teredo* has a very small shell, although short, strong spines are often present. The grouping of Recent genera here is based on the soft anatomy.

Pitar Eocene–Recent: Worldwide

Pitar: beak and hinge
showing teeth

Medium-sized. Valves very convex and similar in shape to *Arctica* (page 119). Beak points forward, **lunule** shallow, **escutcheon** absent. Teeth as shown; front, side teeth well developed, usually three central **teeth** (a) in each valve. Ligamental notch behind beak, otherwise margins closed; lower margin smooth. Pallial sinus present. Ornament of weak concentric ridges.

Neocrassina Jurassic–Cretaceous: E Af

Medium-sized; shallowly convex to thick. Beak points forward, and front part of shell, much smaller than back. Large lunule and escutcheon clearly defined. Ornament of concentric ridges. Two central teeth on each valve. Margins smooth in younger specimens developing small crenulations later in life, perhaps reflecting change from male to female. Margins closed.

Pholadomya Triassic–Recent: Worldwide

Medium-sized to large, elongate. Strongly inflated with two

biconvex valves. Beak near front end, not strong, rounded and pointing upward. Ornament of radiating ridges over central region but with concentric ridges prominent at front and back ends. Teeth absent or weak. Valves with strong back gape and smaller front gape. Pallial sinus present.

Sanguinolites *Devonian–Permian: Worldwide*
Medium-sized. Elongate and curved with front end very reduced. Thick. Teeth absent from dorsal margin; escutcheon large and clearly defined; lunule less well defined. Ornament of concentric ribs. Margins smooth and leaving small gape at back end.

Trigonia *Triassic–Cretaceous: Worldwide*
Left valve shown. Medium-sized to large. Almost triangular with front edge steeper than back. **Beak** pointing upward or slightly backward. Flattened face at back of shell delimited by high ridge and smooth channel. Strongly ornamented with either concentric ridges or radially directed ribs. Escutcheon large and defined by a high crest with a beaded edge. Large central **tooth** (c) on left valve, and two large teeth (d) on right valve, have strongly grooved surfaces. Margins closed and smooth.

Trigonia: beak showing teeth on left (top) and right valves

Schizodus *Carboniferous–Permian: Worldwide*
Small to medium-sized. Thick with flattened margins. **Beak** strong, pointing upward and front end reduced. Lunule and escutcheon absent. Single large **tooth** on each valve (a), a few smaller teeth also present. Shell surface smooth or with weak concentric ripples. Margins smooth and closed.

Schizodus: beak and hinge showing tooth

2in

Pholadomya

Trigonia

Schizodus

Sanguinolites

121

Modiomorpha

Anodonta

Carbonicola

2in

Anodonta (swan mussel)
Cretaceous–Recent (freshwater): NA SA E Af Asia
1–6 inches long. Shell elongate, beak well-formed pointing forward or upward. Shell flattened to thick. Surface smooth or with concentric rings. Hinge toothless or with small ridges. Ridge runs backward from the beak to the back margin. Back margin more pointed than front. Margins closed.

Carbonicola *Carboniferous: E (non-marine)*
Freshwater, medium-sized, flattened to thick. Equal-sized valves, subtriangular in shape. Elongate at back, shortened at front. Beak pointing upward or forward. Dorsal margin curved. Sometimes one or two tooth-like structures present under beak on each valve. Margins smooth, closed. Front muscle scar circular and deep; back scar shallow and high. Ornament of concentric lines.

Modiomorpha *Silurian–Devonian: NA E Asia*
Medium-sized. Equivalve and valves expanded backward. Beak low. Single tooth on left valve and socket on right. Margins smooth and closed. Ornament of concentric lines.

Arca *Tertiary–Recent: Worldwide*
Medium-sized, usually 2–4 inches long. Elongate with beak well in front of midline and pointing slightly forward. Valves very convex. Dorsal margin carrying very wide, flattened areas which separate the beaks. Hinge with long row of small, comb-like teeth. Lower margin often characterized by a wide gape. Ornament of concentric and radial ribs.

Parallelodon *Devonian–Jurassic. Worldwide*
Usually 2–6 inches long. Elongate with very long back region and shortened front end. Beak pointing forward. Dorsal margin straight. Large flattened areas between beaks carry longitudinal ridges. Very few teeth near back of hinge (a), and numerous shorter, curving teeth near front (b). Elongate gape on lower margin. Margins smooth. Externally the shell is marked by strong concentric growth lines and a radial ribbing.

Parallelodon: beak and hinge showing teeth

Glycymeris *Cretaceous–Recent: Worldwide*
Small to medium-sized, almost circular. Beak almost centrally placed and pointing upward (equilateral). Teeth like *Arca* but arranged in gentle curve. Areas developed but smaller than in *Arca*. Crenulations on lower margin. Surface smooth or with radial ridges and concentric grooves.

Modiolus *Triassic–Recent: Worldwide*
Medium-sized to large, up to 4 inches long. Generally similar to the common mussel *Mytilus*, but beak not at the very front of shell. Dorsal margin without teeth. Shell surface smooth or with shallow concentric ridges. Equivalve with ligamental notch developed, otherwise margins smooth and closed.

2in

Dorsal (umbonal) view of left valve of *Arca*

Arca

Parallelodon

Glycymeris

Modiolus

123

Atrina

Pterinopecten

Inoceramus

2in

Gervillella

Inoceramus: beak and hinge
showing ligamental pits

Atrina (pen shell) *Eocene: Worldwide*
Medium-sized to large. Shaped like a half-closed fan, triangular
and up to 10 inches long. Valves equal with beaks at anterior
point. Ornament of wide ripples below and radiating ridges
above. Shiny, inner shell layer surface often exposed (shown
here). Lower margins with elongate gape near front. Back margins wide open. *(Pinna, Jurassic–Recent: Worldwide, is a similar
and related genus is more elongate and differs internally from
Atrina.)*

Gervillella *Triassic–Cretaceous: Worldwide*
Medium-sized to large, up to 10 inches long. Very elongate
with greatly lengthened back and reduced, sharply pointed
front. Dentition of a few elongate teeth which are almost parallel to the long axis. Region above dorsal margin flattened
with numerous (up to ten) vertical pits which hold the ligament. Ornament of concentric growth lamellae.

Inoceramus *Jurassic–Cretaceous: Worldwide*
Medium-sized to large, usually 3–6 inches high. Back wing
expanded as shown or reduced. Numerous **ligamental pits**
(a) along upper edge of dorsal margin and wing. Hinge
without teeth. Ornament of concentric, coarse ripples and
fine grooves. Beak points upward. Shell short and high,
very convex.

Pterinopecten *Silurian–Carboniferous: Worldwide*
Medium-sized. Beak pointing upward. Dorsal margin straight
with wings developed before and behind beak; back wing

2in

Oxytoma

Meleagrinella

larger. Right valve usually less convex than left. Ornament of radial ridges of variable strength.

Oxytoma *Triassic–Paleocene: Worldwide*
Small to medium-sized. Beak pointing upward with wings developed before and behind. Back wing usually longer and pointed. Right valve flattened, left valve convex. Hinge lacking teeth but with narrow areas, that of the left valve continuing in plane of margin and that of the right valve is at about 90° to this. Ornament of coarse ridges with wide intervals. Ridges produced as spines around margin.

Meleagrinella *Triassic–Jurassic: Worldwide*
Small to medium-sized. Small wings before and behind beak. Hinge lacking teeth. Left valve convex, right valve flattened. Left valve with radial ridges which have spiny edges; ridges weak or absent on right valve. A block with many small specimens is shown with mainly left valves visible.

Chlamys (scallop) *Triassic–Recent: Worldwide*
Medium-sized, rarely more than 6 inches high. Similar to living, common scallops. Equilateral, inequivalve, left valve more convex than right. Wings before and behind beak; front wing notched on right valve. Hinge teeth absent but triangular ligamental notch developed under center of beak on both valves. Ornament of strong ribs giving serrated edges at the margins. Concentric ornament is also usually developed. (*Chlamys* is one of several genera loosely referred to as "scallops.")

Chlamys
(right valve)

125

Gryphaea *Triassic–Jurassic: Worldwide*
Medium-sized to large, up to 6 inches long. Left valve much larger than right and very convex with beak rolled over onto right valve and displaced slightly backward. Right valve flat or concave. Ornament of left valve has numerous well-defined **lamellae**. Right valve with smooth or rippled surface and lamellae near margin. Left valve with elongate, curved swelling along back edge above margin.

Actinostreon *Jurassic: Worldwide*
Usually medium-sized. Valves convex, shape varying from similar to *Ostrea* to inequilateral (shown here). Almost equivalve. Radial ridges characteristic, varying from strong ripples to high ridges (shown here); these give lower margin zigzag contact. Inner faces with small tubercles near margins.

Ostrea (common oyster) *Paleocene–Recent: Worldwide*
Medium-sized to large, up to 8 inches long. Left valve moderately convex; right valve slightly smaller than left and flat. Shape varies from circular to more height than length. Both valves have a layered appearance. Left valve with irregular rounded ribs, crossing lamellae. Right valve unribbed with lamellae.

Cardiola

2in

2in

Ostrea

Gryphaea

Actinostreon

Spondylus *Jurassic–Recent: Worldwide*
Medium-sized to large, up to 6 inches high. Nearly equilateral, strongly inequivalve. Valves high, and right valve deeper than left. Dorsal margin straight. **Beak** of right valve with large area (c) which carries fine vertical and cross-striations. Area (f) of left valve low and sloping outward. The specimen shown here is particularly spiny but in some forms the ridges predominate with only a few spines present. Two large **teeth** (e) are far apart on the left valve and close together on the right valve. Deep **notch** (d) below center of beak on both valves.

Spondylus: beak and hinge of left (top) and right valves

Plagiostoma *Triassic–Cretaceous: Worldwide*
Medium-sized to large, up to 6 inches long. Valves same size. Beak points upward and the front edge is straight with an elongate, wide lunule. Margins usually closed. Teeth weak or absent. Surface smooth with fine concentric or radial striations.

Cardiola *Silurian–Devonian: NA E*
Shell small. Beak points upward or forward, equivalve. Hinge teeth obscure. Triangular areas on both valves. Margins may have a gape. Strong radial ribs crossed by concentric grooves give a squared pattern.

Nucula *Cretaceous–Recent: Worldwide*
Shell small, equivalve. Beak points backward. Comb-like teeth along margins before and behind beak (shown here). Internal ligamental process under beak. Lower margin has fine striations. Anterior and posterior muscle scars equal in size. Outer surface may possess a fine ribbed ornamentation and concentric growth rings. Inner surface shiny (shown here).

2in

Nucula

Spondylus

Plagiostoma

127

ARTHROPODS

(Phylum *Arthropoda*)

The largest phylum of animals. It includes insects, spiders, scorpions, crustaceans, millipedes, centipedes, and several extinct groups – of which the **trilobites** (pages 129 to 136) are the most important and the earliest known arthropods. Trilobites were well established by the start of the Cambrian and the dominant arthropod during the Paleozoic.

The most characteristic feature of arthropods is the **tough exoskeleton**, which is slightly flexible in most cases and provides attachment for the muscles. This hard, outer coating protects them from many predators and enables them to survive the harshest of conditions. It also increases their potential for fossilization, and the records of both insects and crustaceans (crabs and lobsters) can be traced back to the Lower Paleozoic. In most arthropods, the body is divided into a **head**, **thorax**, and **abdomen**, with the jointed legs attached to the thorax.

With the exception of trilobites, the arthropods are relatively uncommon as fossils, though insects and crustaceans may be locally abundant. The arthropod groups are extremely large and it is possible to show only a few representatives of the phylum here.

Crotalocephalus gibbus. This trilobite lived from the Lower Ordovician to the Upper Devonian period (around 390–360 million years ago)

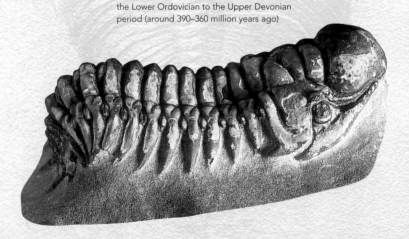

TRILOBITES

The most common fossil arthropods. The trilobite's body divides transversely into the **head** (a), **thorax** (b), and **tail** (c). It is divided along its length by two furrows delimiting the central **axis** (d) from the side regions. The axial region of the head or **glabella** (e) is flanked on each side by the **genae** or **genal regions** (f). The sides of the thorax and tail are known as **pleural lobes** (g).

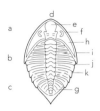

Eyes may be present on either side of the glabella (shown in *Phacops* on page 130 and also on page 12). The rear outer corner of each genal region is termed the **genal angle** (h), which may project as a **genal spine** (i) – shown in *Dalmanites* on this page). A **front border**, a raised rim around the front of the glabella and genae, also may be present.

(shown in *Phacops* on page 130 and also on page 12)

Typical structure of a trilobite as shown by *Dalmanites*

The thorax consists of **segments** defined by **thoracic grooves** (j). The number of thoracic grooves is an important species identifier. The side region of each segment is a **pleuron**. A **pleural furrow** (k) is a groove sometimes present on the upper face of each pleuron.

The **tail** also shows segmentation, and transverse furrows may be present on the axis. The tail is known technically as the **pygidium**, but this term is not used here.

The undersurface of the trilobite is only rarely exposed, but a large plate, the **hypostome**, may be locally very common. This comes from the underside of the head, and probably covered the foregut, behind which lay the mouth.

Trilobite with ramshorns

2in

Dalmanites

Ogygopsis

Dalmanites *Silurian: Worldwide*
Medium-sized. Tail about same size as head. **Glabella** with deep grooves, widening forward; eyes prominent; front border wide; **genal spines** long. Thorax about 11 segments; pleural furrows marked. Tail about 11 segments; back border smooth, carrying a long spine. Ornament of small **tubercles**. (Shown on page 129)

Phacops *Silurian–Devonian: Worldwide*
Head larger than tail. Head dominated by the glabella, which is inflated and strongly convex; eyes large, lenses visible here; front border convex and bounded by deep groove; genal angles rounded. Thorax about 11 segments. Back edge rounded and smooth. Many are found rolled into a tight ball.

Ogygopsis *Cambrian: NA*
Medium-sized to large. Elongate; tail larger than head. Glabella parallel-sided with faint cross-grooves; eyes long and narrow; front border wide and flattened; genal spines short (not shown here). Thorax about eight segments; axis strong and wide; pleurae with deep wide pleural furrows. Tail about ten segments; axis tapering; tail **pleurae** with deep segmental grooves and furrows; back edge with convex border and smooth outline. (Shown on page 129)

Calymene *Silurian–Devonian: Worldwide*
Well-known, medium-sized trilobite. Head much larger than tail. Head semicircular with strongly convex glabella which tapers anteriorly. It is lobed with largest lobes toward the rear. Facial sutures pass from front edge of head around the back of the small eyes to bisect genal angle. Thorax of about 13 segments. Tail fused with six distinct rings on axial region.

Phacops

2in

Calymene

Paradoxides

2in

Paradoxides *Cambrian: NA E Af Aust*

2in

Head much larger than tail. Glabella expanded forward, carrying about three pairs of cross-furrows; eyes large; genal spines about half body length. Thorax of about 18 segments; **pleural furrows** strong and diagonal; pleurae produced as spines at sides, which increase in size backward. Tail small with straight, back edge.

Paedeumias *Cambrian: NA E Asia*

Head large with flattened cheeks. Glabella deeply furrowed. Rounded swelling at front of glabella is connected to front border by a ridge; genal spines long. Thorax of about 14 segments, decreasing in size backward from the second; first segment with short spine; second larger segment with spine extending back beyond tail region; other pleurae with long spines; pleural furrows deep. Tail small, carrying long spine (twisted to left in the specimen shown here).

Olenoides *Cambrian: NA SA Asia*

Head and tail about same size. Glabella with several furrows, expanding slightly forward and reaching front border; eyes medium-sized; front border convex and wide; genal spines short. Thorax about seven segments; axis wide and tapering with cross-furrows and tubercles or spines on each segment; pleurae produced as short spines. Tail of at least five segments with axis tapering backward; back edge with several pairs of spines.

Calymene
(curling stages)

Olenoides

Paedeumias

2in

2in

Encrinurus   *Leonaspis*   *Cheirurus*

Oryctocephalus

Oryctocephalus *Cambrian: NA SA E Asia*
Tail and head almost equal in size. **Glabella** parallel-sided with three or four pairs of cross-furrows which have deep pits at each end; eyes small; **genal spines** long (not clearly shown here). Thorax of about seven segments; **pleurae** produced as spines; **pleural furrows** deep and diagonal. Tail axis with six cross-grooves; sides and back of tail produced as long spines (not clearly shown here).

Encrinurus *Silurian: Worldwide*
Head larger than tail. Inflated glabella widening forward and covered in small pustules, give a rough texture; eyes pronounced; genal spines small, directed outwards. Thorax of 11 or 12 segments. Tail of five to ten pleural segments; back edge serrated. Ornament of three rows of strong **tubercles**.

Cheirurus *Silurian–Devonian: Worldwide*
Tail smaller than head. Glabella produced forward to over-hang front border; eyes medium-sized; genal spines small. Thorax of about eleven segments; pleural furrows short and diagonal. Tail with well-defined, deeply grooved axis; back edge with three pairs of spines separated by small central spine.

Leonaspis *Silurian–Devonian: NA SA E*
Head very wide; eyes large; front border with strong spines; genal spine large (broken off in the specimen shown here). Thorax of about eleven segments; pleurae produced back-wards as spines. Tail small; back edge with one pair of large spines and two pairs of smaller spines.

Cedaria

Triplagnostus

Triplagnostus *Cambrian: NA E Asia Aust*
Small, less than 1 cm. Head and tail same size. Glabella divided
into triangular front and elongate hind lobes, less convex than
in *Eodiscus*; cheeks curved and divided at front by a groove;
front border strong and convex; eyes absent; genal angle
rounded or with small genal spine. Thorax of two segments.
Tail very similar to head; axis of tail slightly wider than glabella,
divided into larger triangular back region and a shorter front
region which may carry a strong swelling. A groove at the back
separates the two curved side regions of the tail. Back border
similar to front border.

Eodiscus (head)

Eodiscus *Cambrian: NA E*
Very small, less than 0.5 cm. Head and tail same size. Head
consists of a short, very convex glabella carrying a single pair
of indistinct furrows, and curved cheek regions which are
divided at the front by a deep groove extending to the
narrow front border; eyes absent; genal angle sharp or
strong; genal spines may be present. Thorax of two or three
segments. Tail axis pronounced and carrying many strong
cross-grooves; tail pleurae swollen and curved. Head and tail
regions are shown separated here – *Eodiscus* is often found
in this condition.

Eodiscus (tail)

Cedaria *Cambrian: NA*
Head and tail almost equal in size. Glabella lacking furrows,
having rounded front end and terminating well behind border; eyes medium-sized; front border strong and convex;
genal spines (not shown here) fairly long. Thorax of about
seven segments; axis well defined by furrows; pleural furrows
long. Tail with strong axis and four or five furrows; back
edge rounded.

2in

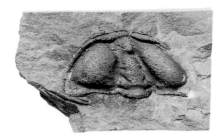

Ctenocephalus (head)

Bonnaspis

Elrathia

Ctenocephalus *Cambrian: NA E Af Asia*

Only head region shown. **Glabella** very convex, tapering forwards, carrying three pairs of strong furrows; cheeks swollen, convex; eyes absent; front border very convex; **genal spines** long, extending over half the length of the thorax (not shown here). Body similar in shape to that of *Elrathia* with small tail and about 15 thoracic segments. Ornament of fine **tubercles** covering head region.

Bonnaspis *Cambrian: NA*

Head slightly larger than tail. Glabella very convex, expanding strongly forwards to front edge; furrows not present on glabella; eyes small; genal spines short (not shown here). Thorax of about seven segments; **pleurae** with deep furrows. Tail up to five segments, poorly defined; back edge rounded.

Elrathia *Cambrian: NA*

Medium-sized. Head much larger than tail. Glabella tapering forwards with rounded front end well behind front border; glabella surface carrying several pairs of weak furrows; eye ridges strong; front border wide; genal spines short. Thorax about 13 segments; **pleural furrows** long and deep. Shallow furrows on tail indicate about five segments, back edge smoothly rounded. *Elrathia* is the most frequently found trilobite in North America.

2in

Trinucleus

Bumastus

2in

Cryptolithus *Ordovician: NA E*

Head much larger than tail. Glabella narrow and very convex, widening forwards and carrying a single pair of furrows; eyes not visible. The most characteristic feature is the wide front border which slopes downwards and outwards, and carries rows of well-developed pits occur around front and sides of head-shield; long genal spines long. Thorax of about six segments. Tail smooth with raised central region and smooth back edge.

Trinucleus *Ordovician E*

Similar in general shape to *Cryptolithus*. Broader than long. Glabella convex and carrying three pairs of deep furrows; fringe of well-developed pits occur around front and sides of head-shield; long genal spines, approximately twice the length of thorax and tail (not shown here). Thorax of six segments; axis strong. Tail much wider than long; back edge smooth.

Bumastus *Ordovician–Silurian: Worldwide*

Unusual elongate with head and tail regions equal in size. Glabella not clearly defined but head carries large swellings on either side; genal angles rounded. Thorax of eight to ten segments; axis not clearly defined. Tail convex with steep back border and smooth outline. Surface ornament very weak.

Cryptolithus

135

Harpes *Devonian: E Af*

The head region only is shown here. **Glabella** very convex with lobes at sides; eyes strong; **genal spines** almost as long as body and very wide; front border wide, carrying many fine pits and tubercles. Thorax of about 29 segments. Tail small.

Griffithides *Carboniferous: NA E*

Medium-sized; elongate. Head and tail almost equal in size. Glabella wide and expanding slightly forward; eyes small; front border narrow; genal angle rounded. Thorax of about nine segments; axis strong. Tail of numerous segments.

Isotelus *Ordovician: NA E Asia*

Head and tail equal in size. Glabella not clearly defined; eyes medium-sized, produced as conical swellings; genal angles rounded. Thorax of eight segments; axis very wide and defined by shallow furrows. **Pleurae** short; pleural furrows short, deep and diagonal. Tail region pointed with weakly defined axial region and weak furrows in the pleural areas.

Harpes

Isotelus

Griffithides

2in

EURYPTERIDS

Eurypterus *Ordovician–Carboniferous: NA E Asia*
An extinct group closely related to the scorpions and important during the Paleozoic. Some eurypterids attained great size, being well over three feet long. Complete specimens are rare but fragments may be locally common. Eurypterids are popularly known as **giant water scorpions** and the largest, *Pterygotus*, was about 10 feet long, and is the largest known arthropod.

CRUSTACEANS

Lobsters, crabs, crayfish, shrimps, prawns, barnacles. One of the most important and diverse groups of marine invertebrates.

Hoploparia *Cretaceous–Eocene: Worldwide*
Small lobster with an elongate but slightly depressed body. Skeleton divided into three distinct regions: head, thorax, and long, segmented abdomen. The **rostrum** over the head is long and narrow. Long, jointed legs and large **chelipeds** (pincers) are typical.

Hoploparia

2in _____

Balanus (barnacles) *Eocene–Recent: Worldwide*
The barnacles or cirripedes are highly specialized crustaceans. Skeleton is modified into protective "shell" composed of large plates. 'Shell' is usually cone-shaped, with a lid. **Opercular** valves, across the opening, are retained in this specimen.

Palaeocarpilius *Upper Eocene: E Af*
Crabs were plentiful during the Eocene and began to resemble modern day species. The example shown, *Palaeocarpilius aquilinus*, lived in seas that covered southern Europe and northern Africa.

Balanus

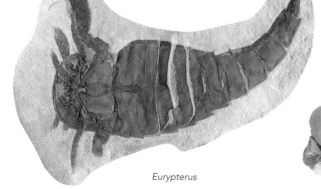

Eurypterus

2in _____

Palaeocarpilius

Acanthochirana *Upper Jurassic: E*
Acanthochirana is a shrimp and belongs to the order Penaeidea. Penaeidea have well-developed rostrums.

2in

INSECTS

Body divided into three parts: head, thorax, and abdomen.
Thorax has three pairs of legs. Wings usually present.

Libellulium *Upper Jurassic–Lower Cretaceous: Europe*
Libellulium is a dragonfly and belongs to the order Odonata.
They are large, predatory insects that first appeared in the
Upper Carboniferous. They have two pairs of equal-sized
wings with a dark spot near the tip, a long abdomen, and
a large head with large eyes and short antennae. True
dragonflies have permanently outstretched wings with veins
that form a distinctive triangle near the base. Fossil forms are
mostly known from incomplete wings, though the Solnhofen
Limestone of Germany is famous for its complete specimens,
as shown here. Wing veins are numerous and are used to
identify fossil species.

Acanthochirana

Anthophorites

Snipe fly *Upper Eocene : Baltic amber*
Amber is fossilized tree resin, produced by trees for self-
defense. It sometimes contains insects that are perfectly preserved. Amber can be purchased from jewelery shops. Most
commercial amber comes from the Baltic region (Upper
Eocene) or the Dominican Republic (Lower Miocene). Flies
(order Diptera, family Rhagionidae) are the most common
insects in Baltic amber and are distinguished from other
insects by having only one pair of wings.

2in

Snipe fly in amber

Libellulium

138

ECHINODERMS

(Phylum *Echillodermata*)

Fossil echinoderms are easily identified because of their characteristic five-fold symmetry and their skeleton of many plates, each of which is composed of a single calcite crystal. Six groups – **sea lilies**, **blastoids**, **starfishes**, **brittle stars**, **edrioasteroids**, and **echinoids** are mentioned here. The sea lilies and sea urchins are the most important as fossils. The echinoderms are varied in form and function, and the overall structure of the skeleton will reflect the mode of life. The starfish and sea urchins are burrowing and free-moving scavengers. In contrast, the sea lilies are mostly fixed sea-floor dwellers: a stem and rooting structure anchoring them to the substrate. A common feature among echinoderms is the presence of an internal complex of tubes and bladders. This is termed the **water vascular system** and may be recognized in living animals as the tubular extensions which aid feeding, movement and respiration. The tubes are called tube feet, and their presence in fossil species is recognized by a sequence of pores in specific plates. Each group of echinoderms has its own characteristic features; these are also important in the identification and classification of the individual specimen.

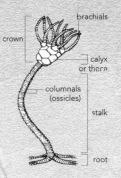

Typical features
of a Crinoid

Key to major groups of fossil echinoderms

1 Skeleton with stem or stalk-like process	▶ 2	
b Skeleton without stem	▶ 4	
2a Skeleton symmetrical	▶ 3	
3a Pentameral symmetry pronounced; small, bud-like cup or calyx consisting of 13 plates	see **blastoids** (page 145)	
b Pentameral symmetry; calyx with regularly arranged rings of plates	see **crinoidea** (page 140)	
4a Skeleton with disk-shaped body and five distinct arms	see **ophiuroidea** (page 144)	
b Skeleton or test rounded; entire, with five pore-bearing areas each two plates wide	see **echinoidea** (page 146)	
c Skeleton cup-shaped with plates arranged in rings around cup	see **crinoidea** (page 140)	

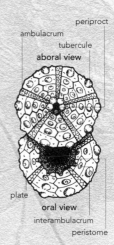

Typical features
of an Echinoid

Conulus

SEA LILIES (Crinoidea)

Plant-like body consisting of a **cup-shaped** body from which arise **five branching arms** which form a filtration cone. Most crinoids also have segmented arms and stems composed of disk-like **ossicles**. Fine side branches from the arms are known as **pinnules**. Often stem ossicles are found in isolation (shown in *Cyathocrinites* on page 142). Some species have become secondarily free-living and lack a stem (shown in *Marsupites* on page 142 and *Uintacrinus* on page 141).

Sagenocrinites Silurian: NA E
Cup large, composed of many hexagonal plates and incorporating the lower parts of the arms. Arms branching dichotomously, without pinnules. Stem circular in cross section.

2in

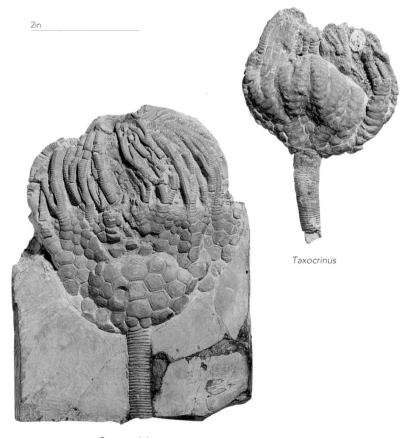

Taxocrinus

Sagenocrinites

Taxocrinus *Devonian–Carboniferous: NA E*
Cup relatively short and including lower parts of arms;
but arm ossicles clearly differentiated from cup plates, the
latter being much smaller than in *Sagenocrinites*. Arms
branching dichotomously, without pinnules. Stem circular in
cross section.

Uintacrinus *Upper Cretaceous: NA E Aust*
Cup large, composed of many small hexagonal plates, with
lower parts of arms bound into cup. Arms branch once and
bear pinnules. No stem.

Pentacrinites *Jurassic: NA E Asia*
Cup tiny, with long, highly branched arms bearing pinnules.
Arms entirely free of cup. Stem very long, composed of
star-shaped ossicles often found in isolation. The stem has
fine lateral tendril-like projections called **cirri**.

Pentacrinites

2in

Uintacrinus

Marsupites *Cretaceous: NA E Asia, Aust, Af*
Cup large, stemless. Large plated cup composed of three
rows of five plates. The base of the stem is occupied by a single large plate. The plates have a pattern of well-defined
ridges. Arms short and rather small.

Cyathocrinites *Silurian–?Permian: Worldwide*
Cup small, bowl-shaped, composed of three circlets of large
plates. Arms well-separated and free of cup, branching
dichotomously; lacking **pinnules**. Stem circular in cross-
section. Several discoidal stem elements can be seen
scattered among the arms and to the top left of the specimen
shown here. These have a wide central hole and a
marginal arrangement of radiating ridges and grooves.

Phanocrinus *Carboniferous: NA E Af*
Cup small and bowl-like, composed of three rows of large,
fused plates. The five pairs of thick, strong arms consist
of columns of rounded **ossicles**. Stem long and circular in
cross section.

2in

Marsupites

Phanocrinus   *Carpocrinus*   *Cyathocrinites*

2in

Platycrinites

Glyptocrinus

Dichocrinus

Platycrinites *Devonian–Permian: NA E Asia*

Cup large, composed of just two circlets of large polygonal plates, the upper with five plates, the lower with three. Arms free of **theca** and branching two or three times close to base; composed of a single series of plates near their base but becoming double (biserial) distally; bearing pinnules. Stem ovate in cross section and characteristically helically twisted.

Glyptocrinus *Ordovician–Silurian: NA*

Cup large, conical, composed of a large number of plates ornamented by strong radial ridges, and incorporating the lower parts of the arms. Arms branching dichotomously twice; bearing pinnules. Stem cylindrical with pentagonal central perforation.

Carpocrinus *Silurian: NA E*

Cup moderately large, conical and incorporating the lower parts of the arms; plates polygonal. Base of **theca** composed of three approximately equal-sized plates (a, b, and c). Arms branching once in cup so that there are ten strong free arms bearing pinnules. Stem circular in cross section.

Dichocrinus *Carboniferous: NA E*

Cup bowl-like, composed of two circlets of large plates, the lower circlet composed of just two plates. Arms generally branched just once and composed of double elements except close to base where they are single. **Pinnules** on arms long and well-developed. Stem circular in cross section. Here *Dichocrinus* lies on top of a specimen of *Rhodocrinus*, which is similar in general appearance but has a cup composed of numerous small plates.

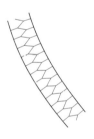

Platycrinites: biserial arm

Carpocrinus: base of theca

Dichocrinus: base of theca

STARFISHES (Asteroidea)

Star-shaped body composed of many small platelets loosely bound together, usually with five arms which are not sharply marked off from the body. Some have large skeletal plates edging the body, called **marginal plates**. Plates running down the midline of each arm on the oral surface are termed **ambulacral plates**. Complete starfishes are rare as fossils, but where conditions are right for their preservation they are commonly abundant.

2in

Mesopalaeaster

Mesopalaeaster Ordovician: NA E
Arms narrow, bounded by a single series of marginal plates. Aboral surface (not illustrated) with distinct rows of stellate plates aligned along the arms. Ambulacral plates block-like without gaps.

Calliderma Cretaceous–Recent: Worldwide
Cushion-shaped with short projecting arms and strong frame composed of a double series of marginal plates. Upper surface covered in tessellated platelets with semiregular arrangement.

Pentasteria Jurassic–Eocene: E
Arms rather straight and subparallel-sided with strong marginal plates forming double series. Central body area rather small; covered in small granular platelets on upper surface. Ambulacral plates slender with large gaps.

BRITTLE STARS (Ophiuroidea)

Star-shaped body with narrow, cylindrical arms clearly separated from a circular disk-like body. Ophiuroids are distinguishable by details of their disk and arm plating, and are not readily identifiable without a microscope. Ophiuroids do not have an anus.

2in

Lapworthura Silurian: E Aust
This brittle-star has a large central disk and robust arms. The mouth is placed centrally within a star-shaped arrangements of small plates. As with *Palaeocoma*, the arms extend over the under surface of the disk toward the mouth. The arms are broad, with long vertical spines.

Calliderma   Pentasteria

Palaeocoma *Jurassic: E*
A typical brittle star with long, flexible, whip-like arms and a small disk-shaped body. Mouth is present on undersurface. Mouth is flanked by five distinct **buccal plates**. On the under-surface, the inner or proximal areas of the arms extend inward to reach the mouth.

EDRIOASTEROIDS (Edrioasteroidea)

Discoidal to subglobular echinoderms that attach directly to hard substrata. Most have only the upper surface plated which comprises an outer marginal ring and a central plated body with five sinuous grooves (ambulacra) radiating from a central mouth.

Edrioaster *Ordovician: NA*
Hemispherical with ambulacra extending to edge (**ambitus**) and curved around body. Remaining plates polygonal and tessellate. Marginal frame hardly differentiated and on underside.

Edrioaster

BLASTOIDS (Blastoidea)

Attached, plant-like fossils with stem, bud-like body, and fine feathery appendages termed **brachioles**. The body has five prominent grooves radiating from an apically positioned opening (the mouth) and is composed of a small number of large polygonal plates, including five V-shaped plates associated with each groove. Usually only the plated body is preserved.

Pentremites

Pentremites *Carboniferous: NA*
Clearly defined radial symmetry. It has five ambulacral rays and five **spiricles** or outlets around the mouth. The calyx (cup) is bud-like and comprises a limited number of plates. Each ambulacram forms a V-shaped depression or groove. The specimen illustrated shows just the bud-like body. *Pentremites* is one of the most common blastoids.

2in

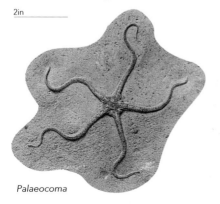

Palaeocoma

Lapworthura

SEA URCHINS (Echinoidea)

Globular, ovate, hemispherical or heart-shaped animals with a
test (body) consisting of numerous plates. These are arranged
into ten radial "segments" that extend around or over the test.
These plates are covered in **spines**, which are only loosely
attached and commonly become detached before fossilization.
Tubercles over the plate surface mark the sites of attachment
of these spines. **Ambulacral plates** are perforated by **pores**
which form easily recognizable radiating tracts. In between the
ambulacra are two columns of **interambulacral plates**. The test
has two openings, the mouth and anus.

Regular echinoids

Echinoid in which the mouth and anus occur in the center of the
adoral and **aboral** surfaces. Tubercles are generally prominent
covering all plates. The structure of tubercles, whether with a
central perforation and whether bearing cog-like crenulation, is
important for identification.

Pedina *Jurassic–Miocene: NA SA E Madagascar*
Subglobular test with rather sparse covering of small tubercles
that are perforate and lack crenulation. Anal opening central
within a circle of five large plates.

Pedina

Psammechinus

2in

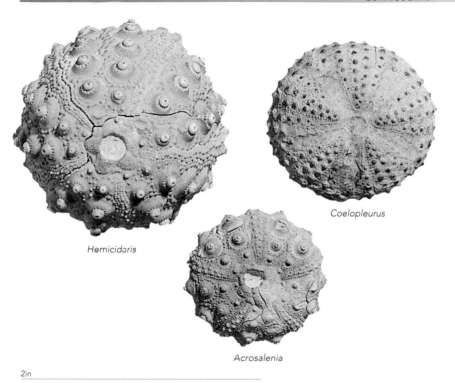

Hemicidaris

Coelopleurus

Acrosalenia

2in

Acrosalenia *Jurassic–Cretaceous: Worldwide*
Small with flattened test. Large tubercles on plates of interambulacral areas. Anus slightly displaced from center of apical area. Two rows of small tubercles occur on each of the ambulacral areas; these are rather sinuous in appearance.

Hemicidaris *Jurassic: NA E Af Asia*
Small to medium sized test. Like *Acrosalenia* but with anal opening central within apical circlet of plates. The interambulacral tubercles are strongly developed and are known to bear spines. Tubercles on ambulacral plates decrease markedly in size on upper surface.

Coelopleurus *Eocene–Recent: Worldwide*
Test depressed with five wide aboral zones free of tubercles. Tubercles imperforate and without crenulation. Pores in single series above, but crowded and forming broad zones toward large mouth opening.

Psammechinus *Pliocene–Recent: NA E Af*
Depressed test covered in rather uniform tubercles which are imperforate and lack crenulation. Plates around anal opening usually lost. Pores offset in arcs of three along the ambulacra.

Irregular echinoids

Anal opening may have migrated from the central position (apex) of the **test**; usually posterior on the oral surface. **Tubercles** generally very fine and uniform. Most have aboral pores double and strongly elongated to form petals.

Pygaster Jurassic: E

Test depressed with large central mouth on underside notched around its edge. Anal opening very large and key-hole-shaped on upper surface, but displaced to the posterior of the apex. Tubercles clearly seen and slightly sunken that support short weak spines. The five **ambulacra** are straight and radiate outwards from the **apical** area.

Micraster Cretaceous–Paleocene: Worldwide

One of the best-known fossil echinoids. Heart-shaped with mouth on lower surface near the base of the frontal groove and anal opening on the posterior surface. Test has slightly inflated appearance with posterior region higher than front. Five petaloid (petal-shaped) arms occur on the upper surface and are indented. Plates at the apex are compact. Two large plates covered in large tubercles form much of the test behind the mouth.

Holaster Cretaceous: Worldwide

Heart-shaped, like *Micraster*, but without sunken petals and with apical plates stretched out along the anterior-posterior axis.

Clypeaster Eocene–Recent: Worldwide

Test very thick-shelled; flattened with a rounded margin. Mouth small and central on lower surface; somewhat sunken with five grooves radiating from it. Rests partially buried on the sea floor, ambulacral areas extend only to margin of test. Petals strongly developed on upper surface. Anal opening small, at posterior margin of lower surface.

Conulus Cretaceous: Worldwide

Conical in profile. Mouth small, circular and central on flat lower surface. Anal opening just below margin at rear of lower surface. Ambulacral pores simple throughout with no petal development. Commonly found as internal casts in flint (shown here) derived from the chalk of Europe.

Pygurus Jurassic–Cretaceous: Worldwide

Test depressed and pentagonal in outline. Small pentagonal mouth opens a little anterior of centre on the lower surface; the anal opening is also on oral surface, at the posterior border. Petals strongly developed aborally.

Echinolampas Eocene–Recent: Worldwide

Test ovate with small pentagonal mouth slightly anterior of center, and anal opening wider than long and visible in oral view at the posterior border. Petals well-developed with unequal length rows of pores in each ambulacrum.

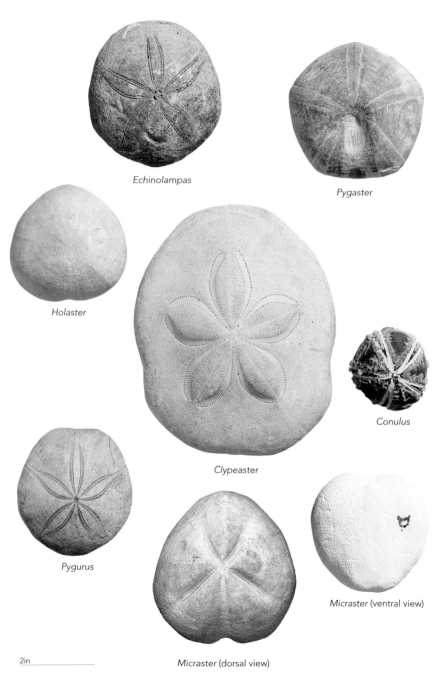

Echinolampas

Pygaster

Holaster

Clypeaster

Conulus

Pygurus

Micraster (ventral view)

2in

Micraster (dorsal view)

GRAPTOLITES

(Class *Graptolithina*)
A group of colonial, usually **planktonic** animals. The class is an extinct group of **hemichordates**, with distant vertebrate affinites. Its members were important and common from the **Cambrian** to the **Carboniferous**. Graptolites are good **zone fossils**. They are important for dating **Paleozoic** rocks because they changed very rapidly through time and many genera had worldwide distribution. Graptolites are common in shales and slates, where they are flattened along the bedding planes and are usually carbonized.

Each graptolite colony is known as a **rhabdosome**, and consists of a variable number of branches or **stipes** that diverge from the initial individual of the colony, which is known as the **sicula**. The **nema** is the thread-like process by which the rhabdosome may be attached. Each individual of the colony is housed in a cup-like structure known as a **theca**.

Often you will only recognize the skeleton as a toothed, blackened branch, but sometimes well-preserved specimens will exhibit a number of diagnostic features. Each graptolite family is characterized by a given number of stipes. The shape and arrangement of the thecae are also important in the recognition of specific genera. Throughout the **Ordovician** and **Silurian** periods, there is a progressive reduction in the number of stipes in successive graptolite families.

Key to major groups of graptolites

Monograptus

1a Stipes numerous; colony shrubby in appearance: show two types of cup under high magnification

example: ***Dendrograptus*** (page 151)

b Stipe number limited to 8 or less

▶ **2**

2a Stipe number 5-8

example: ***Dichograptus***

b 4 stipes

example: ***Tetragraptus*** (page 152)

c 2 stipes

example: ***Didymograptus*** (page 151)

d 1 stipe

▶ **3**

3a Thecae on both sides of stipe

example: ***Diplograptus*** (page 151)

b Thecae on one side of stipe

example: ***Monograptus*** (page 152)

Dendrograptus *Cambrian–Carboniferous: Worldwide*
An attached, plant-like form. Rhabdosome consisting of numerous stipes which give it a fern-like appearance.

Diplograptus *Lower Ordovician–Lower Silurian: Worldwide*
Member of the graptoloid group (Graptoloidea) of the graptolites, which also includes *Monograptus*, *Dicellograptus*, and *Tetragraptus* (see below) Members of this group were important planktonic forms in the Ordovician and Silurian. *Diplograptus* has thecae arranged on each side of the stipes as shown (that is, it is **biserial**). This shows as double serrations on the ribbon-like specimens.

Diplograptus: biserial scandent growth form

Didymograptus *Lower Ordovician: Worldwide*
The various species referred to this genus have two stipes. These may hang down like the two prongs of a tuning fork, or spread outward to form an almost straight line. The **thecae** are usually simple, tooth-like or slightly curved. Two important "tuning-fork" species occur in Lower Ordovician rocks. The most common of these is *D.murchisoni*, which is much larger than the related species *D.bifidus*.

Didymograptus

Diplograptus

2in

Dendrograptus

151

Monograptus: monoserial growth form

Tetragraptus: reclined form (top), pendent form (below)

Dicellograptus: reclined growth form

Monograptus *Lower Silurian–Lower Devonian: Worldwide*
Thecae arranged in a single row along the side of the **stipe** as shown (that is, it is **monoserial**). Rhabdosomes may be coiled, spiral or straight, and fragmentary stipes of the other genera in the family (Monograptidae) also resemble *Monograptus*.

Tetragraptus *Lower Ordovician: Worldwide*
Rhabdosome made of four stipes which branch in pairs, as shown. Each branch has thecae on one side only. Each stipe is characterized by the presence of numerous closely packed, tooth-like thecae. Serrated edges are clearly shown on this specimen.

Dicellograptus *Lower–Upper Ordovician: Worldwide*
Consisting of two stipes that are characteristically flexed from the center as shown, and carry thecae on one side only.

Monograptus

Dicellograptus

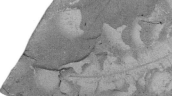

Tetragraptus (reclined form)

2in

VERTEBRATES

Fishes, amphibians, reptiles, birds and mammals are all vertebrates. Vertebrates have internal skeletons of cartilage or bone. Some, like early fish, turtles and armadillos, have a heavy protective armor which may either cover the head or encase the body. Complete fossil skeletons are rare; isolated bones or teeth being much more common.

For a more detailed understanding of fossil vertebrates, you should consult specialist books and visit a museum, as one animal may be known by hundreds of bones. Sharks can be readily identified by their sharp pointed teeth, and individual mammals on the shape and size of individual canine or cheek teeth. Fish bones are often shiny brown or black in color, although no hard and fast rule exists for easy identification. Books such as *Palaeozoic, Mesozoic and Cainozoic Fossils*, published by the British Museum (Natural History), will help with the identification of the more common discoveries. It is recommended, however, that large, intriguing and important specimens are reported to your local museum. At the museum, the curators of the vertebrate collection will be able to help in the determination of your material.

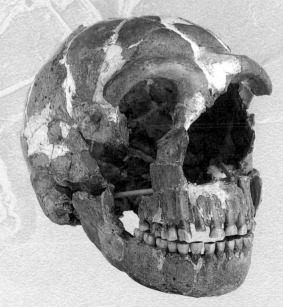

Skull of *Homo neanderthalensis*

FISHES

The largest group of living vertebrates with more than 20,000 species and a huge number of fossil forms.

Armored fishes

Many Paleozoic fishes had a heavy external armor of bone. These are usually found isolated. Fossils are usually of Silurian and Devonian age.

Cephalaspis *Silurian–Devonian: NA E Asia*

One of the best-known armored fishes. A complete specimen is shown. The head of this genus is covered by heavy bony shield, and the body by thick scales. Prominent "horns" or spines extend backward from the corners of the headshield. The eyes occur in the center of the dorsal surface.

Coccosteus *Middle and Upper Devonian: Eu Asia NA*

A skull roof is shown; shield-shaped, dorso-ventrally compressed, and composed of paired and median bony plates, joined at sutures and ornamented with tubercles. The lateral plates at the back of the skull roof are modified to articulate with the bony trunk shield.

2in

Cephalaspis

Coccosteus
(skull roof in a concretion)

154

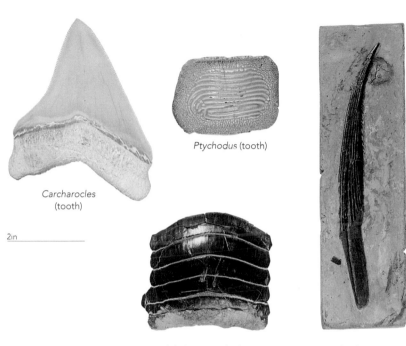

Carcharocles
(tooth)

Ptychodus (tooth)

2in

Myliobatis (tooth plate)

Hybodus (spine)

Sharks and rays

Skeleton composed of cartilage or bone and rarely fossilized. Teeth are quite common in the Carboniferous, becoming more common in the Cretaceous and Tertiary when cartilaginous fish diversified.

Hybodus *Triassic–Cretaceous: Worldwide*
Teeth low and wide, high central point and numerous side points. Spine long and pointed with grooved sides. These spines support the dorsal fins in members of the hybodont group of sharks which were common during the Mesozoic.

Carcharocles *Paleocene–Pleistocene: Worldwide*
Very large teeth with a single point and serrated edges.

Ptychodus *Cretaceous: NA E Af Asia*
Flattened teeth suitable for crushing mollusk shells. The teeth are rather square with a strongly ridged upper surface. This is a hybodont shark, but its teeth are similar to those of many rays.

Myliobatis *Cretaceous–Recent: Worldwide*
The eagle rays, like the sharks, have a cartilaginous skeleton. The teeth are wide and flattened with several teeth joined to form a tooth plate.

Bony fishes

Include most living fishes such as the salmon, cod, and herring. The group was important in freshwater by the end of the Paleozoic, and has since become important in marine conditions. Identification is very difficult.

Cheiracanthus

Middle Devonian: Europe; Lower or Middle Devonian: NA

An articulated fish; body covered with tiny scales, smooth, or ornamented with faint longitudinal ridges. A long, slender fin spine supports the anterior margin of each fin. The shoulder girdle is preserved here, just above the base of the pectoral fin spine, behind the head.

2in

Cheiracanthus
(in a concretion – apparent
spines are preparation marks)

20in

Lepidotus

2in

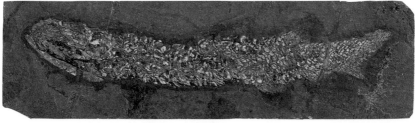

Osteolepis

Brookvalia (on a bedding plane – apparent spines are preparation marks)

Perleidus (in a nodule)

Ceratodus (tooth plate)

2in

Osteolepis *Middle Devonian: Europe*
An articulated, lobe-finned fish. The scales are rhombic. The two dorsal fins and the anal fin are each separate. The paired fins had a short, fleshy lobe at the base. All fins have an internal skeleton at the base and slender, bony, straight dermal rays (lepidotrichia). The tail is heterocercal.

Lepidotus *Mesozoic: E Af NA*
This genus has a short mouth, lined by strong teeth. The tail is symmetrical, and the body scales thick and shiny with a characteristic diamond shape. The scales are the most common fossils associated with bony fish.

Ceratodus *Triassic–Paleocene: Worldwide*
Lungfish, generally known from fossil teeth only. Shape and ridges are characteristic. Surface with many small pores.

Perleidus *Triassic: E Af Asia*
Complete bony fishes may be found in nodules and the presence of the dead fish sometimes appears to have caused the formation of the nodule.

Brookvalia *Triassic: Aust*
Fossil fishes may be found flattened out along bedding planes and are discovered when the rock is split.

157

2in

Trionyx (carapace plate)

REPTILES

Turtles, ichthyosaurs, plesiosaurs, lizards, snakes, crocodiles, pterosaurs and dinosaurs are all reptiles. Reptiles were the dominant land animals from the Permian to the end of the Cretaceous and the top predators in Jurassic and Cretaceous seas.

Mesosaurs
Late Carboniferous–Early Permian: SA Africa
One of the first aquatic reptiles, resembles a small crocodile. *Mesosaurus* was lightly built with an elongated head and flattened tail – probably used for swimming. Pictured here is *Mesosaurus tenuidens*. Their numerous thin, fine teeth suggest that they were filter feeders.

Crocodiles
Triassic–Recent: Worldwide
Crocodiles are among the commonest fossil reptiles, but they are very difficult to identify generically. Crocodile teeth vary greatly along the jaw of the same individual. They usually have short, sharply pointed crowns (the black upper part), and long roots. Shown here is *Stenosaurus*, an early Jurassic marine crocodile from the Upper Lias, Germany.

Turtles
Triassic–Recent: Worldwide
Pieces of turtle shell or carapace are the commonest parts found. A plate from the upper part of the shell of *Trionyx*, a freshwater softshell turtle, is shown here. In freshwater turtles the plates sometimes have patterns on their upper surfaces, but in marine turtles the surfaces of the plates are smooth. In life, the plates have a horny covering.

2in

Turtle eggs

Mesosaurus

20in

Ichthyosaurus

Ichthyosaurs

Triassic–Cretaceous: Worldwide
Rarer than crocodiles or turtles but important marine reptiles in the Mesozoic, especially the Jurassic. The most frequently found parts are the centra of the vertebrae. The two swellings at the top indicate where the neural arch was broken off. The pictured *Icthyosaurus* has been preserved with the broken-up skeletons of three unborn young inside, a fourth may have just been born.

Plesiosaurs

Jurassic–Cretaceous: Worldwide
Isolated bones and complete skeletons of plesiosaurs are locally abundant in Jurassic and Cretaceous clays and shales. Plesiosaur vertebrate are usually larger and flatter than those of ichthyosaurs. The teeth are longer and pointed but most lack the vertical grooves present in ichthyosaur teeth.

Dinosaurs

Triassic–Cretaceous: Worldwide
Dinosaurs have been found on every continent. Their remains are most abundant in North America, China, and Mongolia. At least 30 different dinosaurs have been discovered in Britain.

20in

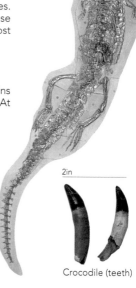

Stenosaurus

2in

Ichthyosaur (vertebra)

2in

Hadrosaur eggs

2in

Crocodile (teeth)

20in

Plesiosaurus

159

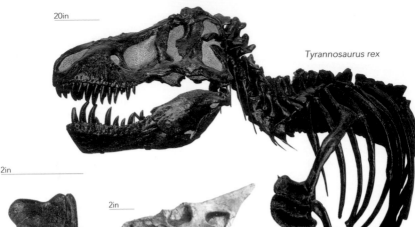

20in

Tyrannosaurus rex

2in

2in

Protoceratops

Tyrannosaurus *Upper Cretaceous: NA Asia*
Large, terrestrial carnivore with six inch (15 cm) -long serrated teeth. It stood at about 40 feet (12 metres). There is currently some discussion among paleontologists whether *Tyrannosaurus* was a predator or scavenger. Shown here is the skull of the most famous of the species – *Tyrannosaurus rex*.

Protoceratops *Upper Cretaceous: Asia*
Herbivorous dinosaur that walked on four legs. *Protoceratops* had a large head with a bony neck frill and parrot-like beak with cheek teeth. The males had larger frills suggesting they were used in courtship displays. The example shown here is a juvenile – the skull could expect to grow to about 46cm (18 in) when adult.

Albertosaurus *Cretaceous: NA*
Flesh-eating dinosaurs have tall sharp, blade-like teeth with serrated edges. A single tooth of *Albertosaurus* is shown here.

Iguanodon *Cretaceous: NA E Asia*
Many plant-eating dinosaurs have square crowned teeth with flat upper surfaces and ridged sides. The skull is shown here.

Hypsilophodon *Cretaceous: E*
Not all dinosaurs were large, and a femur of *Hypsilophodon* is shown here. This plant-eating dinosaur was about 3 feet metre tall and 7 to ten feet long.

Hypsilophodon
(femur)

2in

Albertosaurus
(tooth)

Archaeopteryx *Jurassic: F*
Considered by many to be the earliest known relative of the bird. *Archaeopteryx* is a flying reptile and combines reptilian traits such as teeth and gastralia (stomach ribs) with bird-like feathers and wings. It is one of the most important fossils ever discovered.

Pterosaur *Upper Triassic– Upper Cretaceous: Worldwide*
Species of flying reptile. Fossils are very rare because of the fragility of Pterosaur bones. There are two main types: the Rhamphorhynchoidea and the Pterodactyloidea. The former has short wing metacarpals and a long tail, the latter has long wing metacarpals and a short tail. Pictured is the *Pterodactylus kochi*, the most common of the fossil pterodactyls. Adults had a wingspan of 20 in (50 cm).

20in

Iguanodon

2in

BIRDS

The oldest-known bird dates from the Jurassic. Many different kinds are known from the Cretaceous, although most modern-type birds appear in the Eocene. Their bones are very fragile as they have thin walls and an empty internal cavity; as a result they are only rarely preserved as fossils. You are most likely to find them in Pleistocene deposits, and they can usually be identified by comparison with living bird bones. Shown here is the metatarsus (long part of the foot) of a Dodo (Pleistocene: Mauritius). The foot's form is characteristic of birds, as there are three articulating surfaces for the toes at the lower end; a feature not found in mammals or reptiles. Other bird bones may sometimes be confused with bones of mammals or reptiles.

Dodo (metatarsus)

20in

2in

Archaeopteryx

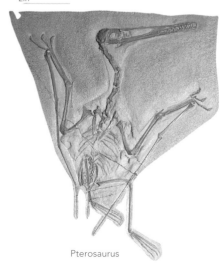

Pterosaurus

161

MAMMALS

Includes the animals which are typically covered with hair and which suckle their young (like humans, horses, elephants, whales, bats, and dogs). The left and right side of the lower jaw are each composed of a single bone, the **dentary**. The teeth socket into the jaw, and are usually well differentiated into distinct functional types (incisors, canines, premolars, molars). A major group of animals since the end of the Cretaceous, mammal remains are common in Pleistocene deposits and locally abundant in some earlier deposits, for example the Oligocene of South Dakota. The teeth are often good indicators of the type of food that was eaten and are very important in the identification of most mammals.

Flesh eaters

Teeth usually modified into either piercing points or shearing blades aligned along the axis of the jaw and are elongate in plan view. The blades often have marked V-shaped notches.

Smilodon (saber-tooth tiger)
Upper Pleistocene: North and South America
Prehistoric member of the cat family (*Felidae*, subfamily *Machairodonontinae*). The long, sharp, backward-curving upper canine teeth of this tiger were adapted to killing large herbivores. Unlike *Smilodon,* other saber-tooth skulls feature a groove on the lower jaw into which the saber-teeth can fit.

2in

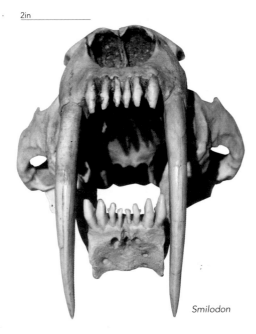

Smilodon

Adcrocuta (skull)

2in

Canis (lower jaw fragment)

Canis (wolf, domestic dog, jackal, dingo)
Miocene–Recent: Worldwide
Dogs have sharp, pointed canine teeth and their cheek teeth have a sharp cutting edge used to slice through flesh and crush bones. Cats, hyenas, weasels and civets also have generally similar slicing teeth.

Adcrocuta (extinct hyaena) *Miocene: E Af Asia*
This is a skull of a young individual, but it shows the long slicing cheek tooth and the relatively small number of teeth. The upper canines are not erupted, but the points are visible near the front of the jaw. The arch of the jaw is wide to accommodate large jaw muscles, and the face is relatively short. These are features of most flesh-eating mammals, but are highly developed in the hyenas which are adapted for crushing bones.

Plant-eating mammals – grazers
Eaters of rougher plant food including grasses. Teeth usually have high crowns, are square to rectangular in plan view and have relatively flat, rough, biting surfaces.

Bos (cattle, including the domestic cow)
Pleistocene–Recent: Alaska E Af Asia
(Transitional between browser and grazer) Upper teeth with four crescentic cusps forming square crown. Lower molars rectangular with an extra cusp at the back of the last molar. Bison (NA), antelopes and gazelles (E Af Asia), deer (world-wide except Aust), and giraffes (E Af Asia) have cheek teeth with similar patterns.

2in

Bos (crown view
of upper molar)

Bos (lower molar)

163

2in

Mammuthus (molar)

Castor (lower jaw)

Mammuthus *Pliocene–Pleistocene: E Af Asia; Recent: Asia*
(Fossil relative of the living Asian and African elephants)
Very large cheek teeth consisting of wide, almost parallel-sided platelets forming ridges on biting surface. Mastodon skulls (*see* page 166) are larger than mammoth skulls and have a flattened brow.

Pliolophus *Eocene: NA E*
Formerly known as *Eohippus* or *Hyracotherium*; *Pliolophus* is the earliest-known horse which stood at 1.6ft (0.5m), about the size of a small dog. Skull long and low. Cheek teeth low crowned with four rounded cusps on the upper molars. You are unlikely to find remains of this animal but it is displayed in most museums.

Castor (beaver) *Pliocene–Recent NA E Asia*
Complete lower jaw shown. Front tooth extremely long with almost triangular cross section and enamel on front face only. Cheek teeth few in number, separated from front tooth by a space, very high crowned, flat-topped with several cross-crests.

Plant-eating mammals – browsers (leaf-eaters)

Consumers of softer plant food and mixed feeders. The teeth usually have quite low crowns and are square to rectangular in plan view. The biting surfaces have well-developed cusps or crests for crushing and shearing.

Ursus *Pliocene–Recent: NA E Asia*
(Transitional between flesh-eater and plant-eater) The genus includes grizzly bears and brown bears. Cheek teeth with low crowns, low rounded cusps and many additional small bumps and grooves. Some pigs have similar cheek teeth. Canines large with swollen root and pointed crown. Diet also includes flesh.

Merycoidodon *Oligocene: NA*
Also known as *Oreodon* and very common in Oligocene of Midwest, USA, where beds are known as "Oreodon beds." Skull relatively short and deep. A leaf-eating mammal having upper molars similar in general crown pattern to *Bos*; consisting of four crescents but crowns much lower. Upper canine relatively large. A sheep-sized relative of camels.

Rhinoceroses (family) *Eocene–Recent: NA E Af Asia*
Upper teeth (upper in plate) with continuous outer walls and two inner crests. Lower teeth (lower in plate) consist of two crescent-shaped ridges.

2in

Ursus (canine)

Ursus (upper molar)

Rhinoceros
(lower molar)

Rhinoceros
(upper molar)

2in

Pliolophus

2in

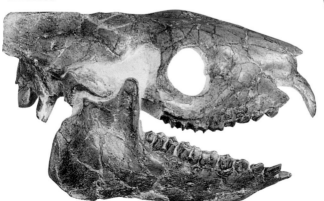

Merycoidodon (skull with lower jaw)

165

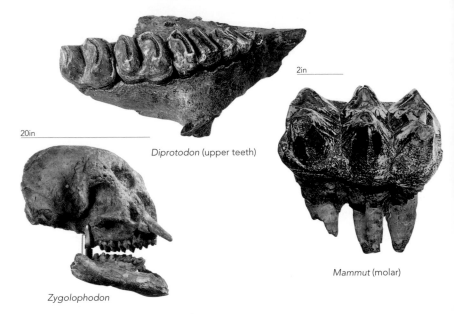

20in

Diprotodon (upper teeth)

2in

Mammut (molar)

Zygolophodon

2in

Hippopotamus
(lower molar)

Diprotodon *Pliocene–Pleistocene: Australia*

Upper teeth shown. These each have a pair of low sharp-edged cross-crests. *Diprotodon* is a marsupial, and is therefore related to the kangaroo, koala, wombat, and opossum. Remains of marsupials are the most common mammalian fossils in Australia. They also occur in SA (Paleocene onwards), NA (beginning in the Cretaceous) and E (Eocene–Miocene). There are rare occurrences in the Tertiary of Af Asia Antarctica. In the Miocene to Pleistocene of E Af and Asia, large teeth similar to *Diprotodon* are from the giant elephant relative, *Deinotherium*, while smaller ones are from pigs or tapirs (not Af). *Pyrotherium* from the Oligocene of SA also had similar teeth.

Mammut *Miocene–Pleistocene: NA E Af Asia*

(Has nothing to do with the mammoth, in spite of its name.) Remains relatively common in North American Pleistocene; known as an American **mastodon**. Cheek teeth large with several cross-crests but these are much lower and more triangular than in the elephantids. The enamel on this type of molar is very thick. The mastodon skull is that of *Zygolophodon*, closely related to *Mammut americanum*. This fossil is perhaps the basis for the myths of the one-eyed giant cyclops.

Hippopotamus *Pliocene–Recent: E Af Asia*

A lower molar shown. Four cusps arranged in a rectangle; similar in general pattern to *Bos* but cusps less crescent-shaped. Some pigs have similar teeth, as do members of an extinct group, the anthracotheres.

Equus (horse, donkey, zebra)
Pliocene–Recent: NA SA E Af Asia
Teeth very high with square crowns (upper in plate) and rect-angular crowns (lower in plate). Pattern complex. In Eocene, for example, *Pliolophus* (shown on page 164), Oligocene and early Miocene horses have low crowned teeth similar to those of small rhinoceroses.

Humans

Although the fossil record is not complete, we know that humans evolved from ape-like creatures. Our earliest ances-tor, *Austrolpithecus afarensis*, lived in northeast Africa some 5 million years ago. Over the next 3–4 million years *A. africanus* evolved. *Homo habilis*, who used primitive stone tools, appeared c.500,000 years later. *H.erectus* is believed to have spread from Africa to regions all over the world from c.750,000 years ago. Records indicate that from *H.erectus* evolved two species, Neanderthal man, who died out 40,000 years ago, and who could have been made extinct by the other species, the earliest modern man *H.sapiens sapiens*.

Homo neanderthalensis *Cretaceous–Oligocene: E Asia*
Known from fossils in Europe and Asia. Bones were thick and powerfully built and the skull had a pronounced brow ridge. Now considered to be seperate species to ourselves, *Homo sapiens*, and possibly a local adaptation during the Ice Ages.

2in

Equus (upper molar)

Equus (lower molar)

Equus: crown view
of upper molar tooth

2in

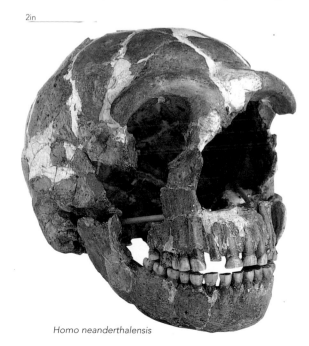

Homo neanderthalensis

FOSSIL LAND PLANTS

Land plants are common fossils, particularly in terrestrial sediments. Plants produce prodigious quantities of seeds, fruits, and pollen or spores, and many species will shed whole organs (e.g. leaves) either continuously or at various times of the year. Many of these plant parts are incorporated into the fossil record. Under certain conditions (e.g. coal swamps), plants are fossilized at their site of growth, providing crucial information on the morphology of the whole plant and important insights into the ecology of ancient terrestrial ecosystems. The fossil record contributes information on climatic and ecological changes as well as data on plant evolution. Pollen and algae are widely used in dating certain types of rocks.

The majority of plant remains are found as **casts** or **molds**, or as **carbonaceous** films on **silicified** lumps of wood. The casts and molds of tree trunks, roots and branches are common in sandstones associated with coal deposits. Impressions of leaves and fronds are frequent in the actual coal scants, along with the carbonaceous remains of the more woody parts. Leaf beds are common in the Mesozoic and Cenozoic, and experts can identify the species by leaf-shape and **venation**. Fossil **forests** are more common in Mesozoic and younger strata, and a detailed analysis of the woody tissues provides a key to identification. A representative selection of fossil land plants is included here. For a more detailed understanding of fossil plants, however, you should consult specialist textbooks and visit a museum.

Araucaria

Zosterophylls
Devonian: NA SA E Asia Aust
Among the earliest known land plants, these fossils are most closely related to living club mosses (lycopsids). They are very simple plants, lacking leaves, roots and seeds.

Sawdonia *Devonian: NA E Asia*
Simple branched stems with coiled tips in younger parts. Stems bearing conspicuous spines.

COAL MEASURES PLANTS

Coal is formed from plants, and the coal measures of the Carboniferous Period are a major source of fossil plants. Spoil heaps at coal mines are excellent places to collect. With the exception of flowering plants, many of the major living groups of land plants had evolved by the Carboniferous Period.

Lycopsids (club mosses)
Late Silurian–Recent: Worldwide
Living lycopsids are relatively small herbaceous plants that are a minor component (< 1%) of modern species diversity. The zenith of lycopsid evolution was the Carboniferous, where as many as 50% of known fossils are attributable to the group. Unlike their living relatives, some extinct species were very large trees up to 30 m (98 ft) in height.

Lepidodendron *Carboniferous–Permian: Worldwide*
One of the best-known coal measure plants. May exceed 76 cm (30 in) in length. Evidence of stems, leaves and roots are all recorded from coal measure sediments. Stem is named *Lepidodendron*, the roots *Stigmaria*, and the leaves and branches *Sigillaria*. Stem is often massive and supports crown of branches. Stem shows diamond-shaped or oval, spirally arranged leaf scars.

Lepidodendron
(leafy branch, *Sigillaria*)

Sawdonia (spiny stems)

2in

Sphenopsids (horsetails)
Devonian–Recent: Worldwide

This group contains only 15 living species, and all are relatively small herbaceous plants. Sphenopsids have a lengthy and diverse fossil record, and many Paleozoic species were large trees up to 20 metres (65 feet) in height. Plants in this group have characteristic jointed stems with branches and leaves in whorls.

Calamites *Carboniferous–Permian: NA SA E Asia Aust*
The stems of living and extinct sphenopsids have a hollow central region. This region may become filled with sediment during fossilization to produce an internal cast of the pith cavity with characteristic vertical ridges and joints.

Annularia *Carboniferous–Permian: E Asia*
Successive whorls of needle-shaped leaves from the terminal branches of the *Calamites* plant. They develop as circlets around the jointed stem and are often preserved as impressions on coal slabs.

Calamites
(internal stem cast)

2in

Annularia

Alethopteris

Ptychocarpus

Neuropteris

2in _____

Fern-like foliage

Devonian–Recent: Worldwide

Pinnate leaves are a common element of Late Paleozoic and Mesozoic floras. Foliage of this type is characteristic of living and fossil ferns as well as extinct seed plants that are more closely related to living cycads, conifers and flowering plants than to true ferns. In the absence of reproductive structures, the precise affinity of much fossil fern-like foliage is difficult to establish.

Pecopteris *Carboniferous–Permian: Worldwide*

Foliage typical of some extinct ferns (e.g. *Psaronius*) and seed plants. **Pinnules** attached along entire width of base; with or without parallel margins; distinct vein extending almost to pinnule tip.

Ptychocarpus *Carboniferous–Permian: Worldwide*

Foliage similar to *Pecopteris* but with spore-bearing organs consisting of numerous microscopic circular structures attached to leaf surface (hand-lens required) that demonstrate an affinity with ferns rather than seed plants.

Neuropteris *Carboniferous: E NA SA Asia*

Foliage of extinct seed plants called medullosans. Difficult to seperate overall shape and form from that of living fern. Leaves are compound and carry many small leaflets. Strong **venation** present, with numerous veins arising from a distinct midrib.

Alethopteris *Silurian–Recent: E NA NAfrica Asia*

Alethopteris is a coal measure seed fern similar to *Neuropteris*. It has a multipinnate leaf and a characteristic venation. The leaflets are slimmer and straighter than those of *Neuropteris*, and the veins shorter and less curved.

Pecopteris

Cordaitanthus (cone)

2in

Cordaites

Cordaitales
Carboniferous–Permian: Worldwide
This extinct group includes the ancestors of the living conifers. Cordaitales were a conspicuous component of the Late Paleozoic flora, and the group included large trees as well as small shrubs.

Cordaites *Carboniferous–Permian: Worldwide*
Fragment of a leaf showing the long, strap-like form characteristic of the group. Veins are parallel to the long axis of the leaf.

Cordaitanthus *Carboniferous–Permian: Worldwide*
This name is applied to either ovulate or pollen-producing cordaitalean cones. These are typically loosely constructed organs. Compare this structure with the more compact cone of *Araucaria* (on page 173).

MESOZOIC AND TERTIARY PLANTS

Ginkgoales
Permian–Recent: Worldwide
This group contains a single living species, *Ginkgo biloba* (SE China), but it was an important and diverse element of Mesozoic floras. The living *Ginkgo* is a large tree.

Ginkgo (maidenhair tree) *Permian–Recent: Worldwide*
Representative of one of oldest groups of nonflowering vascular plants. The characteristic fan-shaped leaves are typically two-lobed in the living species and many fossils. The leaves of some extinct species have many more lobes. Leaves have simple dichotomous veins. They occur in clusters at the end of short branches.

Coniferales (conifers)
Triassic–Recent: Worldwide
An important living group of seed plants that includes pines and redwoods. Leaves are usually long and narrow, and seeds are borne in cones. Conifers diversified during the Triassic, and they were a major component of Jurassic and Cretaceous floras.

Araucaria *Triassic–Recent: Worldwide*
Living Araucariaceae are confined to the southern hemisphere and comprise about 40 species. *Araucaria* includes the monkey-puzzle tree and the Norfolk Island pine. Silicified fossils of *A.mirabilis* cone are shown here with helical pattern of scales, and polished section of cone with bract-bearing ovules.

2in

Sequoiadendron (giant Sequoia) *Tertiary–Recent: NA*
From California, the giant sequoia (Taxodiaceae) is one of the largest living trees. Various species are known through their wood and their cones. Bears small fossil cones with relatively few scales.

Sequoiadendron
(cone)

Ginkgo   Araucaria   Araucaria

173

Bennettitales

Triassic–Cretaceous: Worldwide

An extinct group with ovules and pollen organs grouped into elaborate, flower-like heads. The group is not closely related to living cycads despite some remarkable superficial similarities in habit and leaf morphology.

Williamsonia *Triassic–Cretaceous: Worldwide*

Robust stem and numerous frond-like leaves. Flower-like cone showing radially arranged petal-like bracts. They are solitary and held pollen on stamens that curved inward and upward.

2in

Pterophyllum *Triassic–Recent: Worldwide*

Bennettitalean leaf resembling some cycads and ferns. The pinnules have parallel margins and veins.

Cycadales (sago palms)

Permian–Recent: Worldwide

Small living group of seed plants comprising ten genera including the sago palm (*Cycas revoluta*). Male and female cones are borne on separate plants. This group is only distantly related to Bennettitales and flowering plants.

Nilssonia *Permian–Cretaceous: Worldwide*

Leaves are lanceolate or pinnate with fine parallel veins. Strongly resembling *Pterophyllum* (pictured left) and distinguishable on microscopic epidermal characters (morphology of stomates).

Pterophyllum

2in

Williamsonia (cone)

Nilssonia

Angiosperms (flowering plants)

Cretaceous–Recent: Worldwide

More than 80% of living land plants (c.250,000 species) are angiosperms. Angiosperms probably originated in the Triassic, but the earliest unequivocal fossils come from the Lower Cretaceous. Evidence from fossil pollen grains and leaves documents a rapid diversification of angiosperms in the Early Cretaceous, and this group is the dominant element of many floras by the Early Tertiary. The most commonly collected fossils are leaves and wood. Minute, charcoalified flowers and seeds can also be found in Cretaceous clays. These represent the remains of plants that have been burned in bush fires before fossilization. Recovery of flowers and seeds usually requires special preparation and equipment. Appropriate clay sediments are disaggregated in water, and the organic material is sieved and examined with a microscope. Some common leaves, fruits and woods are listed below.

Dicotyledons

Most flowering plants are dicotyledons, and the earliest fossil angiosperms belong to this group. Primitive living dicotyledons include Nymphaeales, Piperales, Aristolochiaceae, Winteraceae, Chloranthaceae, Calycanthaceae, Laurales, and Magnoliales. Several living groups are recognizable by the Late Cretaceous (Turonian), including Lauraceae, and fossils that are probably related to Chloranthaceae, Magnoliaceae, platanoids, and rosiids. Leaves of dicotyledons typically have a complex network of veins.

Laurus (laurel) *Tertiary–Recent: E Af*

Leaf is long and its edge undivided or entire. A strong central rib is present and the secondary veins diverge from it. Living Lauraceae comprise some 3000 species of mainly tropical and subtropical trees and shrubs. This family has an extensive Tertiary fossil record and has been documented in the Cretaceous.

2in

Laurus

2in

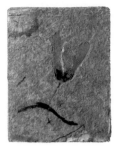

Acer seeds

2in

Platanus (plane) *Tertiary–Recent: Asia E NA*
Palmately lobed leaf. Living Platanaceae contain approximately eight species of temperate and tropical trees. Leaves, wood and **infructescences** are first documented in the mid-Cretaceous. This family is an important component of angiosperm floras throughout the Late Cretaceous and Early Tertiary.

Zelkova (caucasian elm) *Oligocene–Recent: E Asia*
Leaf with one tooth per secondary vein. Living *Zelkova* (Ulmaceae) comprises six or seven species of tree. The group has an excellent European fossil record.

Rhus (varnish tree) *Tertiary–Recent: NA Asia*
Rhus is a member of the Anacardiaceae which comprise some 600 living species. Family includes the cashew and pistachio.

Acer (maple, sycamore) *Tertiary–Recent: Asia E NA Af*
Broad three-lobed leaf with serrated margin. The central rib present in all three lobes; a network of smaller veins branch out over the surface. There are some 140 living species of Aceraceae, most of which are trees or shrubs. Characteristic winged fruits, seeds and leaves are common in the Paleocene and Oligocene. Leaves of *Acerites* are known from the Cretaceous.

Populus (poplar) *Tertiary–Recent: NA E Asia*
Ovate leaves with a crenate margin. Flowers are borne in a raceme. Poplar (and willow) is in the Salicaceae, a group containing some 536 species. Fossils resembling living Salicaceae are known from the middle Eocene.

Acer

Zelkova

Rhus

Populus

2in

Platanus

Fossil wood

Wood is frequently preserved in the fossil record through replacement by minerals such as silicates, calcium and magnesium carbonates, and pyrite. Because of the robust nature of the plant cell wall, soft-tissue preservation at the cellular level is much more common in plants than animals. Special techniques are required to cut and polish fossil wood, and observation of cell structure requires a microscope. Gymnosperm wood (conifers and their relatives) is generally homogeneous with long straight tracheid cells, and it lacks large vessel cells. Angiosperm wood is more heterogeneous, and usually contains **tracheids** and large vessels as well as one or more categories of fibers. Because of the attractive patterns caused by mineralization and the structure of the wood itself, polished sections of fossilized wood are frequently sold in rock shops.

2in ____

Quercus (oak) *Tertiary–Recent: NA E Asia Af*
There are some 450 living species of oaks (Fagaceae), which are widely distributed in temperate and tropical regions. The polished section through this silicified trunk shows conspicuous growth rings. Growth rings in fossil wood can provide important information on paleoclimate. The presence of rings indicates growth in a seasonal climate (e.g. temperate), while the absence of rings suggests a nonseasonal climate (e.g. humid tropics).

Fossil fruits

Fruits are abundant in the fossil record. Soft tissue preservation at the cellular level is a common feature and can provide much information on the affinity of the fruit. Pollen attached to the stigmatic surface can sometimes be used to link fruits to particular fossil flowers.

Quercus (polished
section through trunk)

Prosopis (mesquite) *Eocene–Recent: NA SA Af Asia*
Part of fruit (legume, Mimoseae) with six seeds visible. Living *Prosopis* comprises some 44 species of nitrogen-fixing trees that live in frost-free, arid environments.

Prosopis

Anonaspermum *Eocene–Recent: Worldwide*
Internal cast (pyrite) of storage tissue (endosperm) of seed is shown here. The endosperm has a characteristic and easily recognizable corrugate or punctate surface. *Anonaspermum* is a member of the Anonaceae, a family of predominantly lowland tropical trees, shrubs, and climbers.

2in ____

Fossil fruit
(*Palaeowetherellia*)

Anonaspermum

177

Monocotyledons

About 22% of living angiosperms are monocotyledons, and over half of this diversity is contained in four families: Orchidaceae (orchids), Poaceae (grasses), Cyperaceae (sedges), and Arecaceae (palms). The Cretaceous fossil record of monocotyledons is poor compared to that of dicotyledons. Evidence of a rapid diversification is provided in the Late Cretaceous by fruits of Zingiberales (gingers and their allies) and the leaves and stems of palms. Many groups of monocotyledons had evolved by the Early Tertiary. Leaves of monocotyledons typically have parallel veins.

Sabalites (palm) *Late Cretaceous–Recent: Worldwide*
Often large fossils, this fragment of palm leaf shows the parallel veins typical of monocotyledons. The veins are visible as ridges running along the leaf.

Nipa (palm) *Tertiary–Recent: Worldwide*
These palm fruits are often large and are common fossils in certain Eocene deposits (e.g. London clay). Most living species of palms are plants of tropical and subtropical climates.

Palmoxylon (palm) *Late Cretaceous–Tertiary: Worldwide*
One of the most common fossil members of the palm family. This silicified stem contains evenly distributed vascular bundles. There are no growth rings in this specimen.

2in

Palmoxylon
(cross section of
stem)

Sabalites

Nipa (fruit)

TRACE FOSSILS

Trace fossils are the result of biological activity on or within a sediment. They take the form of tracks and trails, burrows and borings, and even body waste deposits. Dinosaur footprints occur as distinct trails in ancient lake-shore sediments, whereas trilobite tracks crisscross sedimentary rocks laid down on Paleozoic sea floors. The shape or geometry of trace fossils varies with the needs of the animals that created them. Single vertical burrows indicate that the main requirement is for protection against predation or a rigorous environment, while complex horizontal patterns may be linked with a shortage of food and the need for an intensive "farming" of the sea floor.

Cruziana *Lower Paleozoic: Worldwide*
The name *Cruziana* is used in reference to the various tracks and trails formed by trilobites and trilobite-like arthropods. A typical track consists of two lobes, which are the result of the animal scratching its way across the sediment. The traces are often long, and the lobes are parallel with a marked median depression. The scratch marks on the lobes were created by the walking part of the branched limb. More than 30 species of *Cruziana* have been identified. The elongate tracks are indicative of direct movement while searching for food or during migration. Heart-shaped traces, again two-lobed and with scratch marks, represent temporary resting sites and were created by free-swimming animals.

Coprolites and fecal pellets *Throughout fossil record: Worldwide*
These are fossilized droppings and body waste materials of animals. The term "fecal pellet" is used in association with the small droppings of gastropods or crustaceans, while coprolite is used to describe the fossilized feces of larger animals such as crocodiles and dinosaurs. Coprolites may exceed 12 in (30 cm) in length or width, whereas fecal pellets rarely exceed 0.4 in (1 cm) in length.

Coprolites

Cruziana

FOSSIL LOCATIONS

NORTH AMERICA

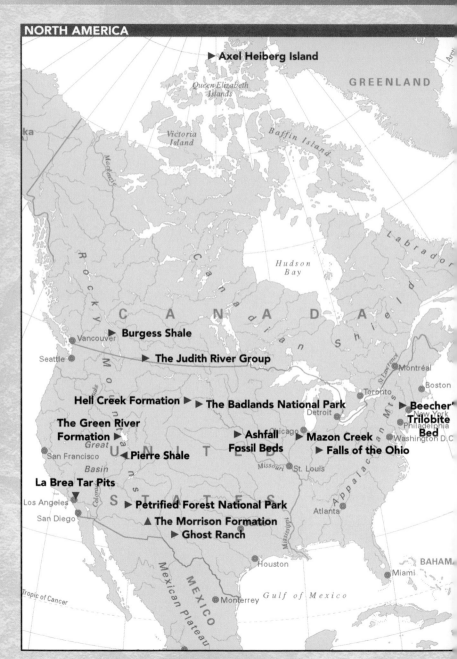

▶ **Axel Heiberg Island**

GREENLAND

Queen Elizabeth Islands

Victoria Island

Baffin Island

ka

Mackenzie

Labrador

Hudson Bay

C A N A D I A N   S h i e l d

R O C K Y   M O U N T A I N S

▶ **Burgess Shale**

Vancouver

Seattle

▶ **The Judith River Group**

Montréal

Boston

Toronto

St. Lawrence

Hell Creek Formation ▶ ◀ ▶ **The Badlands National Park**

Detroit

New York

▶ **Beecher' Trilobite Bed**

The Green River Formation ▶

Chicago

Philadelphia

Washington D.C.

▶ **Ashfall Fossil Beds**

▶ **Mazon Creek**

A p p a l a c h i a n   M t s

San Francisco

U N I T E D

Great Basin

◀ **Pierre Shale**

Missouri

St. Louis

▶ **Falls of the Ohio**

Colorado

La Brea Tar Pits

S T A T E S

Los Angeles

San Diego

▶ **Petrified Forest National Park**

Atlanta

▲ **The Morrison Formation**

▶ **Ghost Ranch**

Mississippi

M E X I C O

Mexican Plateau

Houston

BAHAM.

Miami

Tropic of Cancer

Monterrey

Gulf of Mexico

AUSTRALIA

Coral Sea

Normanton
Cairns
Charters Towers
Townsville
► **Riversleigh**
Mount Isa
Mackay
Winton
QUEENSLAND
Rockhampton
Gladstone
● Birdsville
Bundaberg
Charleville
Maryborough
Gympie
Quilpie
Brisbane
Cunnamulla
Toowoomba
Lake Eyre
Ipswich
● Marree
Lismore
Lake Torrens
Broken Hill
Tamworth
► Coffs Harbour
Flinders Range
NEW
Port Macquarie
Dubbo
SOUTH WALES
► **Ediacara Hills**
Orange
Lithgow
● Newcastle
● Adelaide
Bathurst
● Sydney
Murrumbidgee
Goulburn
Wagga Wagga
Wollongong
Canberra
Shellharbour
Albury
► **Naracoorte**
VICTORIA
Australian Alps
Mount Gambier
Ballarat
Geelong
Melbourne
► **Dinosaur Cove**
King I.
Bass Strait
Tasman
Flinders I.
Furneaux

BRITAIN

► **The Rhynie Cherts**
Aberdeen
SCOTLAND
Dundee
North Sea
Glasgow
Edinburgh
Southern Upland
Newcastle-upon-Tyne
LAKE DISTRICT
Douglas
Isle of Man
York
Belfast
Blackpool
Leeds
Irish Sea
Liverpool
Manchester
► **Dukershaw/ Westhoughton**
► **Wenlock Limestone**
Birmingham
E N G L A N D
Norwich
ANGLIA
WALES
Cotswolds
Oxford
Swansea
London
► **Wealden Formation**
Bristol
Southampton
EXMOOR
Eastbourne
Exeter
► **Jurassic Coast**
DARTMOOR

GERMANY

Koblenz
HESSE
Thuringian Forest
Ore M
Wiesbaden
Frankfurt
Bayreuth
► **Budenbach**
Mainz
Darmstadt
Würzburg
OURG
PFALZ
Mannheim
RLAND
Ludwigshafen
Heidelberg
Nuremberg
Karlsruhe
Heilbronn
Regensb
BAVARIA
Pforzheim
Stuttgart
ICE
BADEN-
► **Solnhofen Limestone**
WÜRTTEMBERG
Ulm
Augsburg
► **Holzmaden**
Freiburg
● **Munich**
Konstanz
Bodensee
A
L
P
S
A U S T
SWITZERLAND
LIECHTENSTEIN

CHINA

● La
B
Guangyuan
Chengdu
Lushan
Zigon
Kunming
▼ **Chengjiang Deposits**
BURMA
VIETN
THAILAND

IMPORTANT FOSSIL LOCATIONS

BRITAIN

Jurassic Coast, *Dorset and Devon*
Situated along the Dorset and East Devon coast of south-west England, the Jurassic Coast is a designated World Heritage Site. The Jurassic coast contains a number of internationally important fossil localities, and an almost complete sequence of Triassic, Jurassic, and Cretaceous rock exposures. Each layer of rock contains numerous **ammonites** (perfect **zone fossils**) which help to identify the time period of the rock strata. From east to west, sites include:
1. **Purbeck Beds**, *Studland Bay, near Swanage, Dorset*
World's most diverse Late Jurassic and Early Cretaceous vertebrate fauna. One of

Fossilized *tree from the fossil forest at Lulworth Cove, Dorset. The donut-shape was formed by algae at the base of the tree.*

the most important sources of fish from the Late Jurassic (more than 30 species found.)
2. **Purbeck** *(including* **Lulworth Cove**)*, Dorset coast*
Unique Late Jurassic fossil forest. The forest grew on the edge of a hypersaline lagoon that, 140 million years ago, covered much of southern England. During that time, algae accumulated around the base of the tree trunks, forming perfect, donut-like fossil rings. Algae also covered huge fallen logs, the fossils of which can be seen today. The waters rose again, submerging the forest and allowing for the trees' excellent preservation. Fossilized soil and pollen has also been found in the area, as well as some exceptionally well preserved wood in silica.
3. **Kimmeridge Bay**, **Portland Harbour**, *and* **Ringstead Bay**, *Dorset*
Kimmeridge Clay is one of the world's richest sources of reptiles from the Upper Jurassic. **Note**: Permission needed
4. **Furzey Cliffs**, *near Weymouth, Dorset*
Oxford Clay is the UK's richest source of Upper Jurassic (Oxfordian) fossils.
5. **Isle of Portland**, *Dorset*
Portland Limestone has yielded the best Late Jurassic marine reptiles, including type specimens of dinosaurs, turtles, ichthyosaurs, and plesiosaurs.
6. **West Bay** and **Burton Bradstock**, *near Bridport, Dorset*
Forest Marble contains a unique assemblage of microvertebrates such as dinosaurs, pterosaurs, fish, crocodiles, amphibians, and mammals.
7. **Charmouth**, *Dorset*
The ammonite *Asteroceros obtusum* is found only at Charmouth.
8. **Lyme Regis**, *Dorset*
World-famous fossil site. Many diverse fossils from the Lower Jurassic. The Liassic limestone contains numerous ammonites. Some of the larger ammonites are preserved in the limestone pavement.

Bickershaw/ Westhoughton,
Greater Manchester, north-west England
Diverse range of **Upper Carboniferous** fossils have been found here, including

Fossil trilobite, Trinucleus abruptus, *from the Ordovician, found in the Wenlock limestone, Builth Wells, Powys, Wales, United Kingdom*

plants, crustaceans, arthropods, fish and coprolites.

Rhynie Cherts, *Aberdeenshire, Scotland*
In 1912, Dr William Mackie discovered the first examples of Lower Devonian land plant fossils at Rhynie Cherts. The finds became the basis for one of the most influential papers on plant fossils. The Rhynie Cherts, now a grassy meadow, has yielded fossils ranging from single-celled organisms to land plants and arthropods.

Wealden Formation – *The Wealds (Surrey, Sussex, and Kent), Isle of Wight, France, and Belgium*
The Wealden Formation has yielded more species of Early Cretaceous dinosaur than any other fossil site. Examples of the dinosaur Iguanodon have been found throughout the region, while fossilized dinosaur footprints have been found in southern England.

Wenlock limestone, *West Midlands and Welsh borders*
Wenlock limestone contains some of the most important and diverse fossils of the Silurian period. The limestones contain many fragmented fossils – the result of sea erosion – and many rare life-assemblages. The area of **Wren's Nest** and **Castle Hill** in **Dudley**, West Midlands, has provided more than 600 species of marine invertebrate, including beds of articulated crinoids and soft-tissue remains. Dudley is a type locality or '*lagerstätten.*' It contains more than 186 species, 63 of which are unique to the area.

UNITED STATES

Falls of the Ohio State Park, *Clarksville, Indiana*
The limestones at the Falls of the Ohio State Park are composed of the remains of numerous Devonian organisms, making it one of the largest naturally exposed Devonian fossil beds in the world.

Mazon Creek, *north-east Illinois*
The Mazon Creek deposit is an example of a '*lagerstätten.*' More than 400 species from at least 130 genera have been identified from Mazon Creek nodules. Both soft- and hard-part fossils have been found.

Morrison Formation
The Morrison formation is a fossil deposit that covers 1.5 million square kilometers of the western United States, stretching from New Mexico in the south to Canada in the north, and from Idaho in the west to Nebraska in the east. The beds are from the Late Jurassic, and have yielded numerous dinosaur discoveries including four species of *Diplodocus.* In 1877 the deposit became the focus for the bitter rivalry between the paleontologists Othniel Marsh and Edward Cope.

Badlands National Park, *North and South Dakota*
The Badlands National Park contains

fossils from the late Eocene and has the world's richest Oligocene fossil beds. Findings at Badlands have helped scientists study the evolution of a number of present-day mammals including the horse, pig, and rhinoceros.

Hell Creek Formation, *North and South Dakota, and Montana*
The formation includes the remains of some of the last dinosaurs. Hell Creek is the only fossil dinosaur bed that crosses the K-T boundary (the period covering the dinosaur extinction).

Pierre Shale, *Colorado and California*
Upper Cretaceous mudstone and shale, famous for its ammonite fossils.

Ashfall Fossil Beds State National Park, *north-east Nebraska*
Hundreds of perfectly preserved articulated skeletons have been discovered in volcanic ash below the farmlands of Nebraska. Examples of rhinos, horses, camels and tortoises have been uncovered, the smallest of which appeared to die in a stream of volcanic lava, and the larger to have died of dust poisoning. The unique environment has also allowed for the preservation of stomach contents and unborn young, providing information about lifestyles and climate.

Rancho La Brea Tar Pits, *Los Angeles, California*
Crude oil has been seeping out of the ground at La Brea for c.40,000 years. Prehistoric animals from the Plesitocene era have been found in pools of asphalt, suggesting that they became stuck while still alive. La Brea contains nearly 650 fossil species. It has the largest assemblage of Ice Age plants and animals. By studying the succession of animals, paleontologists have been able to work out a great deal about the evolution of modern-day mammals, and the lives of extinct animals such as the saber-toothed tiger.

Fossilized remains of trees in the Petrified Forest National Park, Arizona, US. Water containing silica seeped through the logs and encased their woody structures in silica.

Beecher's Trilobite Bed, *Rome, New York*

Brilliantly preserved fossils of Ordovician trilobites have been found in a small quarry near Rome, New York. The trilobites are rare in that their limbs and muscles are still intact. Species include *Cryptolithus bellulus* and *Cornuproetus beecheri*.

Green River Formation, *western Colorado, eastern Utah and south-western Wyoming*

During the Eocene, a number of large inland lakes covered the region, burying the many fossilized organisms found today. Invertebrate fossils are abundant, and 60 vertebrate taxa have been found including fish, reptiles, birds, and mammals. Perhaps the world's oldest bat, *Icaronycteris index*, was unearthed at Green River, complete with stomach contents and cartilage.

Ghost Ranch, *near Abaquiu, New Mexico*

Hundreds of skeletons of the dinosaur *Coelophysis* have been discovered at Ghost Ranch. The finds range from young juveniles to adults of various strength and size.

Petrified Forest National Park, *north-east Arizona*

The Petrified Forest National Park is situated in the middle of the Painted Desert of Arizona. The park features one of the world's largest and most colorful concentrations of petrified wood. The fossil-bearing rocks contain the bones of Late Triassic reptiles and amphibians.

CANADA

Axel Heiberg Island, *Arctic, Canada*

The island holds some of the most unusual fossil plant examples in the world. Rather than being replaced by silica, the Eocene trees and fauna have retained their original carbon-based material. They are so well-preserved that the wood can still be burned. The trees were protected against silicification by silt in the floodwaters.

Burgess Shale, *Yoho National Park, British Columbia, western Canada*

Burgess Shale is a Cambrian rock formation in the Canadian Rockies. Fossils were first discovered in 1909 by Charles D. Walcott, then Secretary of the Smithsonian Insitution. Burgess Shale fossils are important because they include extremely rare, soft-part fossils, which provide important information about the development of early organisms. The Smithsonian's National Museum of Natural History currently house more than 65,000 specimens.

Judith River Group, *Alberta, Canada*

The group comprises three fossil locations: the Foremost, Oldman, and Dinosaur Park Formations. The fossils date from the Cretaceous and vary from microorganisms to whole dinosaur discoveries. Royal Tyrell Museum, Drumheller, is the world's largest collection of dinosaur fossils.

AUSTRALIA

Ediacara Hills, *north of Adelaide, South Australia*

The fossils found in the Ediacara hills date back to the Precambrian era, placing them among the oldest fossils in the world. The hills gave their name to the soft-bodied fauna, the first diverse and well-preserved Precambrian assemblage to be studied in detail.

Riversleigh, *north-west Queensland*, and **Naracoorte**, *South Australia*

These two sites provide exceptional evidence of the evolution of Australian wildlife. Riversleigh provides the first records for many Australian mammals, including the marsupial mole and feather-tailed possum. It also contains fossils of now extinct mammals such as the marsupial lion. Naracoote fossils cover the faunal changes over several ice ages. Fossils range from frogs to buffalo-sized marsupials.

Dinosaur Cove, *Otway range, Victoria, south-east Australia*
Dinosaur Cove and Lightning Ridge are the two leading dinosaur sites in Australia. The rock containing the fossils is very hard, and dinosaur finds have mostly been made by mining companies. Examples include the holotype of *Laeallynasaura*, and a small dinosaur called *Timimus*.

GERMANY

Solnhofen Limestone, *Bavaria, Germany*
Fossils are quite rare in the Solnhofen Limestone of Germany, but discoveries are often exceptionally detailed. As well as vertebrate and invertebrate animals, the limestone also contains soft part fossils. The Solnhofen limestone revealed that Archaeopteryx had feathers.

Holzmaden, *Württemberg, Germany*
Sea fossils from the Jurassic period are common at Holzmaden, the most impressive of which are on display in the Hauff Museum, Holzmaden. Fossils include many famous examples of Ichthyosaur.

Bundenbach, *near Birkenfeld, west Germany*
The Hansrück slate (*Hansrückerschiefer*) contains a diverse fauna of small, Lower Devonian fossils such as brachiopods, trilobites, and coral. To date, more than 260 species have been described from the Hunsrück Slate, of which more than 60 are crinoids. The fossils are usually covered with a thin layer of pyrite.

OTHER

Chengjiang Deposits, *near Kunming, Yunnan Province, south-west China*
Contains Burgess Shale-like fossils dated from the Lower Cambrian. Many soft-part fossils have been recovered.

FOSSIL CODES

Countries have different rules regarding fossil collecting. In the United States, there is controversy surrounding a recent law that makes it an offense to collect fossils from federal land. This is controversial because amateur collectors are often responsible for uncovering important fossils. Even if the fossil is not significant to the paleontology field, many believe that amateurs actually help to conserve fossils that would otherwise be eroded through time. However, an increased trade in fossils has also meant that many important fossils are kept in private collection, passing from dealer to dealer for large amounts of money, and remaining largely hidden from the scientific world.

Whatever your personal stance on the fossil-collecting debate, it is important that you read through the rules regarding fossil collection wherever you intend to search, and obtain the appropriate permissions.

The Paleontological Society in the United States issued its own Code of Fossil Collecting, the main points of which are detailed below.

THE PALEONTOLOGICAL SOCIETY
The Code
1. The principal importance of fossils is for scientific, scholarly, and educational use of both professionals and amateurs.
2. The numbers of specimens of fossils vary widely but certain fossils in all taxonomic groups are rare, and that conserving and making available for study significant fossils and their contextual data is critical.
3. To leave fossils uncollected assures their degradation and ultimate loss to the scientific and educational world through natural processes of weathering and erosion.
4. Prior notification will be made and permission or appropriate permits will be secured from landowners or managers of private or public lands where fossils are to be collected.
5. All collections will be in compliance with

federal, tribal in the case of Native American lands, state, and municipal laws and regulations applied to fossil collecting. 6. The collector(s) will make every effort to have fossil specimens of unique, rare, or exceptional value to the scientific community deposited in or sold to an appropriate institution that will provide for the care, curation, and study of the fossil material.

ENGLISH NATURE
Code of Good Practice
Access and ownership Permission to enter private land and collect fossils must always be gained and local bylaws should be obeyed. A clear agreement should be made over the future ownership of any fossils collected.

Collecting In general, collect only a few representative specimens and obtain these from fallen or loose material. Detailed scientific study will require collection of fossils in situ.

Site management Avoid disturbance to wildlife, and do not leave the site in an untidy or dangerous condition for those who follow.

Recording and curation Always record precisely the locality at which fossils are found and, if collected in situ, record relevant horizon details. Ensure that records can be directly related to the specimens concerned. Where necessary, seek specialist advice on specimen identification and care (e.g. from local museums.) Fossils of key scientific importance should be placed in a suitable repository, normally a museum with adequate curatorship and storage.

Sites of Special Scientific Interest SSSIs are designated by **English Nature** as a legal protection for important wildlife spots or geological features. Important fossil sites are often SSSIs, in which case you will need to gain permission from the landowner before

The most famous *of all Archaeopteryx specimens, found in the limestones of the late Jurassic at Solnhofen, Germany, in 1877.*

being allowed on the site. To find out information about a specific SSI – contact a local team.

It is an offence under Section 28P of the Wildlife and Countryside Act 1981 (as incorporated by the Countryside and Rights of Way Act 2000), without reasonable excuse, intentionally or recklessly to destroy or damage any of the flora, fauna, or geological or physiographical features by reason of which land is of special interest, or intentionally or recklessly to disturb any of those fauna. A person found guilty of any such offence may be liable on summary conviction to a fine not exceeding £20,000 or on conviction on indictment to a fine.

GLOSSARY

aboral Upper surface of sea-urchin test, on which the anus is found.

adoral Lower surface of sea-urchin test, on which the mouth is found.

alveolus Small cavity. Houses cone-shaped part of belemnite shell.

ambulacral plates Any of five radial bands on oral surface of echinoderm.

aperture Opening to outside (shells).

autozooecia (autopores) Larger tubes in a bryozoan skeleton.

bifurcate Branched rib on surface of shell.

bioclast Organic matter in limestone.

biserial Distribution of **thecae** on both sides of graptolite stipe or branch.

brachioles Feathery appendages of blastoids.

calice Outer, youngest, or oral end of a corallite.

camera(e) Chambers in shell. Chambers are separated by partitions (**septa**) and filled with gas.

chitin Skeleton or test of organic, horny substance of calcium carbonate or sand grains.

clast Fragment of rock (lithoclast) or fossil (**bioclast**) material found in sedimentary rock.

columella Central column of a shell.

columellar plications Ridges on columella.

corallite Single starlike radial structure of coral.

crenulate Edge divided into small, tightly folded shape. Ridges on bivalves are crenulations.

cystopores Cystlike calcification.

denticulate Toothlike appearance.

dissepiment Small, often curved vertical plate that develops inside boundary wall of coral.

dorsal valve One of two valves which make up shell of brachiopod; bears support structure for internal feeding organ.

epitheca Outer wall of coral skeleton.

escutcheon Flattened depression behind beak of bivalve.

evolute Shell in which successive coils are in contact, but do not overlap.

facies Distinctive set of characteristics that occur within a given rock, e.g. grain size or texture.

foramen Bounded opening found at or near beak of ventral valve of many brachiopods.

genae Flanking axial region of trilobite head.

genal angle Rear outer corner of each genal region of a trilobite. May project as a genal spine.

glabella Axial region of trilobite head.

guard Dense, bullet-like structure, also called the rostrum.

inter-radial Plates between "rays" of crinoid cup.

involute Shell in which coils are overlapped.

keel Ridge-like feature found on outer whorl of some ammonoid shells.

lacunae Spaces between the ranks in corals.

lamella(e) Thin sheet of calcite or aragonite, characteristic of various bivalve shells.

ligament notch Depressed area along the hinge line of bivalves. Houses ligament.

lonsdaleoid Blistery dissepiments in corals.

lunarium Hooded structure to apertures.

lunule Depression in front of bivalve beak.

mesozooecia (mesopores) Smaller tubes in Bryozoan skeleton.

nema Thread-like structure by which rhabdosome may be attached.

neural arch Arched structure of bone. Occurs above the central mass of vertebra.

oncolite Banded blast (fragment) of algal origin.

operculum Lid-like structure.

orthocone Straight, slender shell.

oscule (osculum) Opening on upper surface of sponge for outlet of water.

pallial line Curving linear mark joining the front and back muscle scars in a shell.

pallial sinus Inflexion of pallial line.

pennulae Small outgrowths on coral **septa**.

periostracum Thick, dark colored, organic layer, often covers aragonite molluscs.

phaceloid branching Single corallite per branch.

phragmacone Chambered portion of a belemnoid skeleton.

pinnules Fine side branches growing from main branches of crinoid.

planispiral Simple type of coiling in gastropods.

pleural lobes Sides on thorax and tail of trilobite.

pleuron Side region of each segment of trilobite.

polyp Soft, flexible body.

pro-ostracum Liplike projection of belemnoid phragmacone.

radial One of five identical plates in crinoid calyx.

ramose branching Branches composed entirely of numerous corallites.

ramp Part above gastropod shoulder.

rhabdosome Complete graptolite colony.

rostrum Dense, bullet-like structure, also called the **guard**, major part of belemnoid skeleton.

rugae Shells with very strong concentric ridges.

septum Vertical calcareous wall or plate found in corals, and in mid-line of certain brachiopods.

shoulder Main angle in whorled gastropods, where shell turns inwards towards suture.

sinus Indentation found along partial line of many bivalve shells.

siphuncle Tube in shell that extends from chamber to initial coil, piercing mid-region of septa.

spicules Thin, rod-like elements support soft body of a sponge.

stipes Graptolite branches.

sulcus Depression found on ventral valve.

suture Line that marks contact between outer wall and internal septum in shells and tests.

tabulae Horizontal element that divides skeleton of various corals; may be curved or flat.

theca Cup-like structure supporting a corallite.

tubercule Small, rounded projection.

umbilicus Hollow centre in the columella.

ventral valve One of two valves which make up shell of brachiopod; houses the ventral.

whorl Complete coil of gastropod shell.

zoarium Skeleton of bryozoan colony as a whole.

zooecium Calcareous tube or box built by individual zooid.

zooid Independent animal body.

INDEX

Number in **bold** type points to major reference
Asterisk (*) next to number indicates illustration